D0580069

THE BRAIN

AN ILLUSTRATED HISTORY OF NEUROSCIENCE

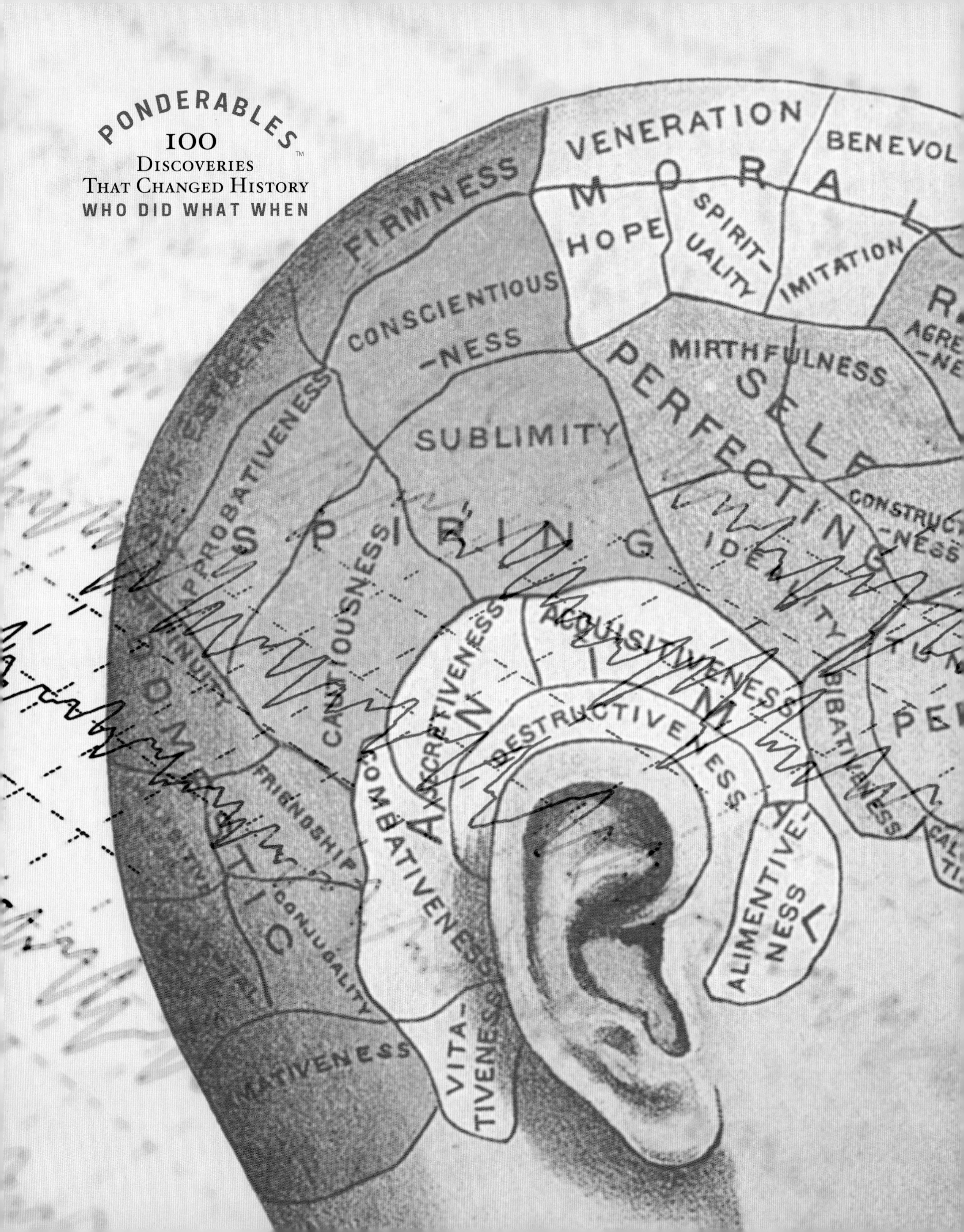
PONDERABLES™
100
Discoveries
That Changed History
WHO DID WHAT WHEN
VENERATION
FIRMNESS
HOPE
SPIRIT-UALITY
IMITATION
CONSCIENTIOUS-NESS
MIRTHFULNESS
SUBLIMITY
APPROBATIVENESS
CAUTIOUSNESS
ACQUISITIVENESS
DESTRUCTIVENESS
COMBATIVENESS
ALIMENTIVENESS
FRIENDSHIP
CONJUGALITY
VITATIVENESS
AMATIVENESS

THE BRAIN

AN ILLUSTRATED HISTORY OF NEUROSCIENCE

Tom Jackson

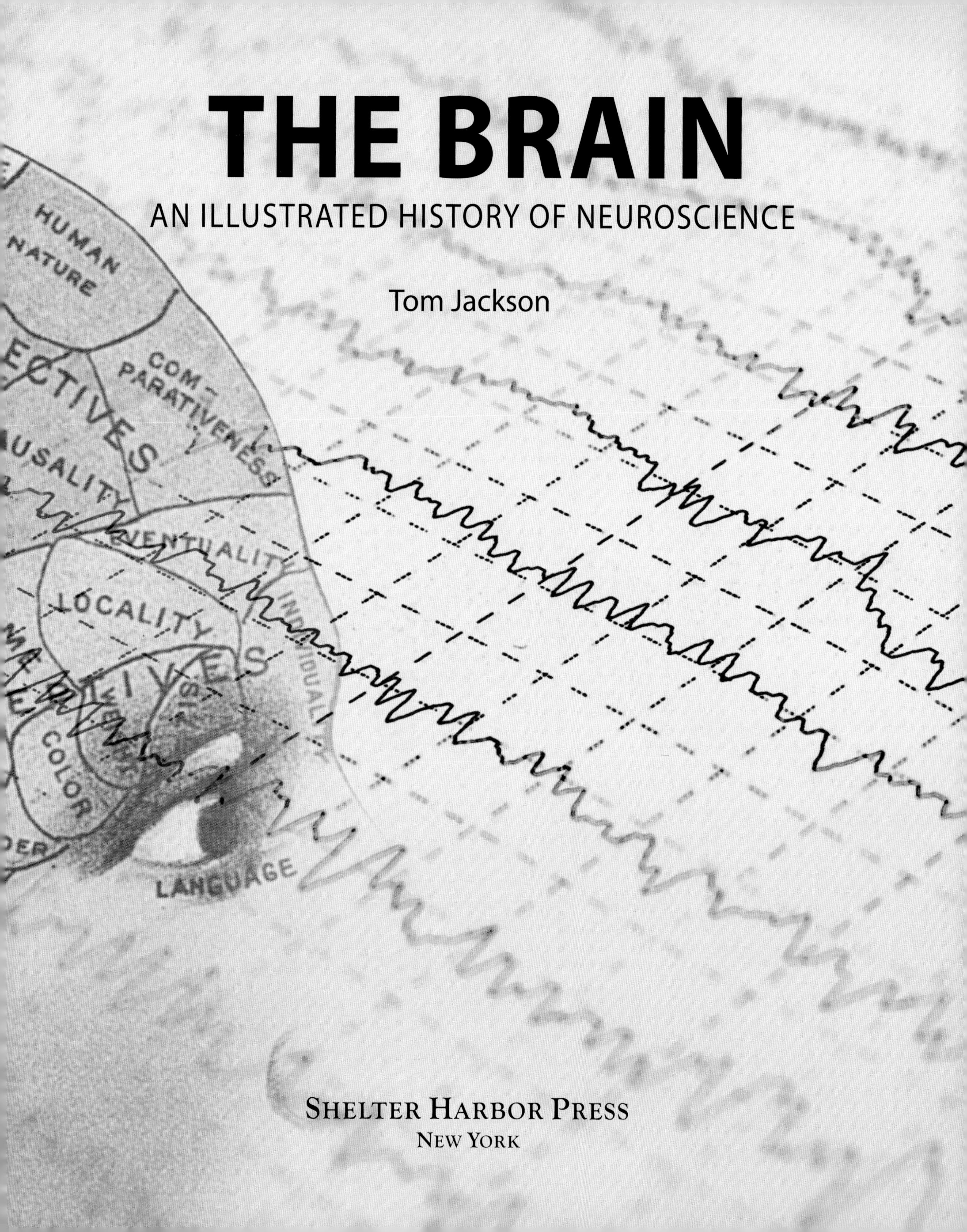

SHELTER HARBOR PRESS
NEW YORK

Contents

INTRODUCTION 6

1 A Hole in the Head 10
2 The Ancient Egyptian Brain 12
3 The Evil Eye 13
4 The Brain in China 13
5 Hippocrates and his Humors 14
6 Theories of Vision 15
7 Three Souls 16
8 Ancient Theories of Sleep 17
9 Galen's Pathways 18
10 The Brain in Section 19
11 Flying Man 20
12 Optical Eyes 20
13 Passions and Emotions 22
14 Dancing Mania 22

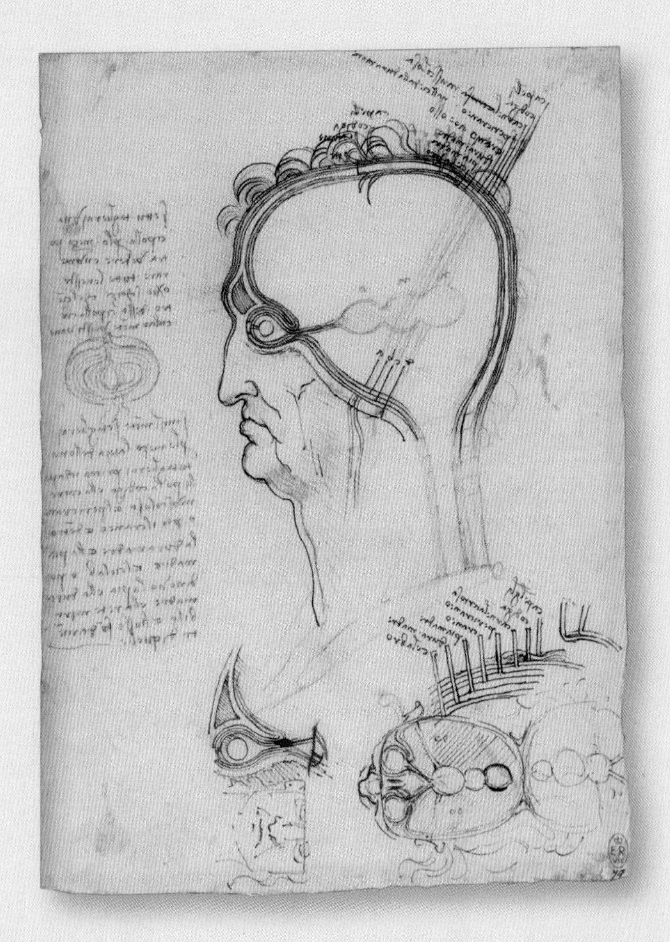

15 Da Vinci's Waxworks 23
16 Michaelangelo's Hidden Brain 24
17 Vesalius's Dissection 25
18 A Witches' Disease 26
19 Apoplexy 26
20 Descartes: Reflex and Reason 28
21 The Circle of Willis 30
22 Functional Anatomy 31
23 St Vitus' Dance 32
24 The Nature of Knowledge 33
25 Idealism 33
26 The Optic Chiasm 34
27 Animal Electricity 34
28 Phrenology 36
29 Parkinson's Disease 38
30 Bell-Magendie Law 39
31 The Neuron 40
32 Anesthetics 42
33 Phineas Gage 44
34 The Neurology of the Ear 45
35 Olfaction 46
36 Glial Cells 47
37 The Speech Center 48
38 Taste Buds 49
39 Neuroscience and Racism 50
40 Electrical Stimultation 52
41 Mood Disorders 53
42 Nerve Nets 54
43 Sensory and Motor Centers 55
44 Phantom Limbs 56
45 Charles Darwin on Emotions 57
46 The Structure of the Eye 58
47 The Black Reaction 60
48 Intentionality 61
49 The Microtome 61
50 Electrical Encephalography 62
51 Hypnotism 63
52 Narcolepsy 64

53 The Visual Cortex 64
54 Tourette's Syndrome 66
55 The James-Lange Theory of Emotion 67
56 Cerebral Dominance 68
57 Psychoanalysis 70
58 Sleep Deprivation 72
59 Whole Brain Functions 72
60 Touch Sensors 73
61 The Synapse 74
62 Autonomic Nervous System 76
63 Bipolar Disorder 77
64 Apraxia: Movement Disorders 78
65 Dementia 78
66 Dyslexia 80
67 A Functional Map 80
68 Sympton Versus Function 82
69 Schizophrenia 82
70 Epilepsy 84
71 Nerve Center: The Striatum 85
72 IQ 86
73 The Cerebellum 88
74 The Gestalt Movement 89
75 Neurotransmitters 90
76 Equipotentiality and Mass Function 91
77 The Hypothalamus 92
78 Theories of Hearing 93
79 Electroconvulsive Therapy 94
80 Lobotomy 95
81 Autism 96
82 Constitutional Psychology 97
83 The Corpus Callosum 98
84 Half a Brain: Hemispatial Neglect 98
85 The Hearing Brain 99
86 Behaviorism 100
87 The Limbic System 101
88 Brain Machines 102
89 Cognitive Behavioral Therapy 103
90 Action Potential 104

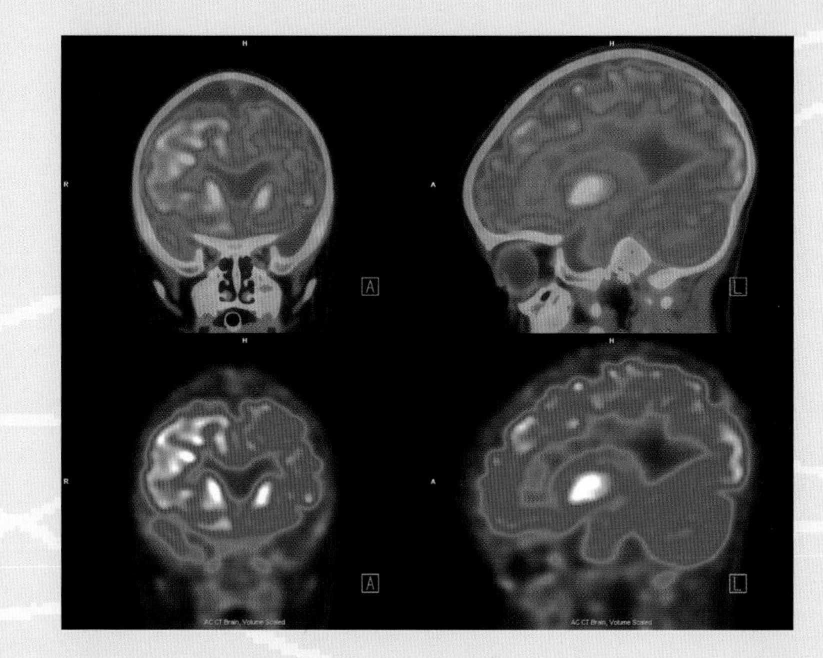

91 The Sleep Cycle 106
92 The Memory Trace 108
93 Coma 109
94 Positron Emission Tomography 110
95 Identity 110
96 Functional MRI 112
97 Parapsychology 113
98 Hard Problems of Consciousness 114
99 Personality or Neurology? 116
100 Computer Brains 117

101: The Brain: the basics 118
Imponderables 124
The Great Neuroscientists 130
Bibliography and Other Resources 140
Index 141
Acknowledgments 144

A Timeline History of Neuroscience Back pocket
Blind Spot Test
Optical Illusions

Introduction

THE HUMAN BRAIN IS THE MOST COMPLEX SYSTEM IN THE UNIVERSE, WITH 83 BILLION NEURONS AND TRILLIONS OF CONNECTIONS. THEN AGAIN THE BRAIN IS NOTHING MORE THAN 3 POUNDS OF FAT AND PROTEIN INSIDE YOUR HEAD. FIGURING OUT HOW ONE MAKES THE OTHER IS CALLED NEUROSCIENCE.

Left: Medieval thinkers began to attribute mental faculties to regions of the brain.

The thoughts and deeds of great scientists always make great stories, and here we have one hundred all together. Each story relates a ponderable, a weighty problem that became a discovery and changed the way we understand the brain—and with each step we learned a little more about ourselves.

The term "neuroscience" is a new one, dating from the 1960s. That might suggest that science has only been concerned with the brain and nerves for less than a century.

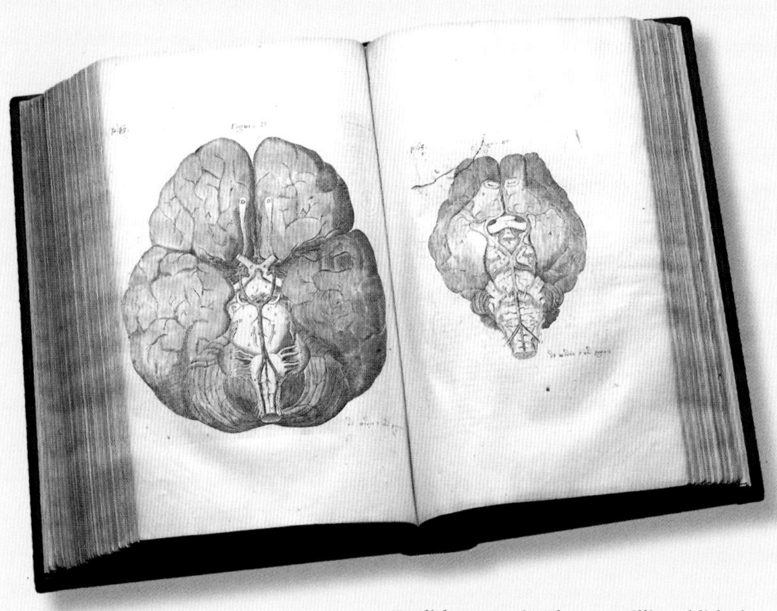

English anatomist Thomas Willis published detailed drawings of the brain's anatomy in 1664, and coined the word neurology.

But while neuroscientists, those dedicated to the working of the brain, are a relatively new breed of researcher, their work was begun long before by neurologists. Neurology is the arm of medical science that concerns disorders of the brain, and dates back in a formal sense to the 17th century, but doctors had been dealing with mental disease and brain injuries for centuries before. It was when the brain went wrong that we first got clues about how it worked.

HEAD OR HEART?

While ancient physicians had some knowledge of how to handle serious head injuries, few cultures put much store in the brain itself. For much of antiquity, the brain was regarded more for its ability to cool the blood than as the control center for the body. That honor belonged to the heart, which was seen as the seat of the emotions and soul (we still refer to it as such today). Progress was slow, but gradually, century by century, the true nature of the brain began to be understood. Scientists began to realize that damage to the brain was linked to paralysis, the loss of speech, blindness, deafness, and personality changes. It was therefore common sense that the brain ruled the body. In fact, "the common sense" was one of the first principles of neuroscience, proposed by the 10th-

Duchenne de Boulogne, an 19th-century French neurologist, short-circuited the nervous system with electric currents, with surprising results!

Jean-Martin Charcot gives a demonstration of hypnosis in the 1870s.

century Islamic scholar Avicenna. He used the term to mean how the brain collected information from the different senses and merged them into one "common" sense.

MAKING SENSE

The brain lies at the center of the nervous system, which includes the spinal cord and a network of nerves that permeates every corner of the body. However, the body's senses—the eyes, ears, nose, etc.—are all involved as well, and the study of how these senses work has provided many a window on the workings of the brain. As the field developed, neurologists began to cut into animal brains to see what happened to their abilities, frequently drawing criticism from the public. Better techniques were needed: Electrical stimulation was less invasive and allowed researchers to work with human subjects. The microscope also opened up a new way to understand the brain in terms of the structure and activities of its cells.

Neurology has close links to psychiatry, the medical specialty that treats mental disorders. To fully understand the way the human brain works—its memory system, its self-awareness and ability to imagine, predict, and plan ahead—the organ needs to be studied when working normally. This became the focus of a new type of researcher: the neuroscientist. These brain detectives have discovered many amazing things, but there are many mysteries that still need to be solved. Let's take a look at what we know and what is left to discover.

Modern medical imaging allows us to watch the brain at work in real time.

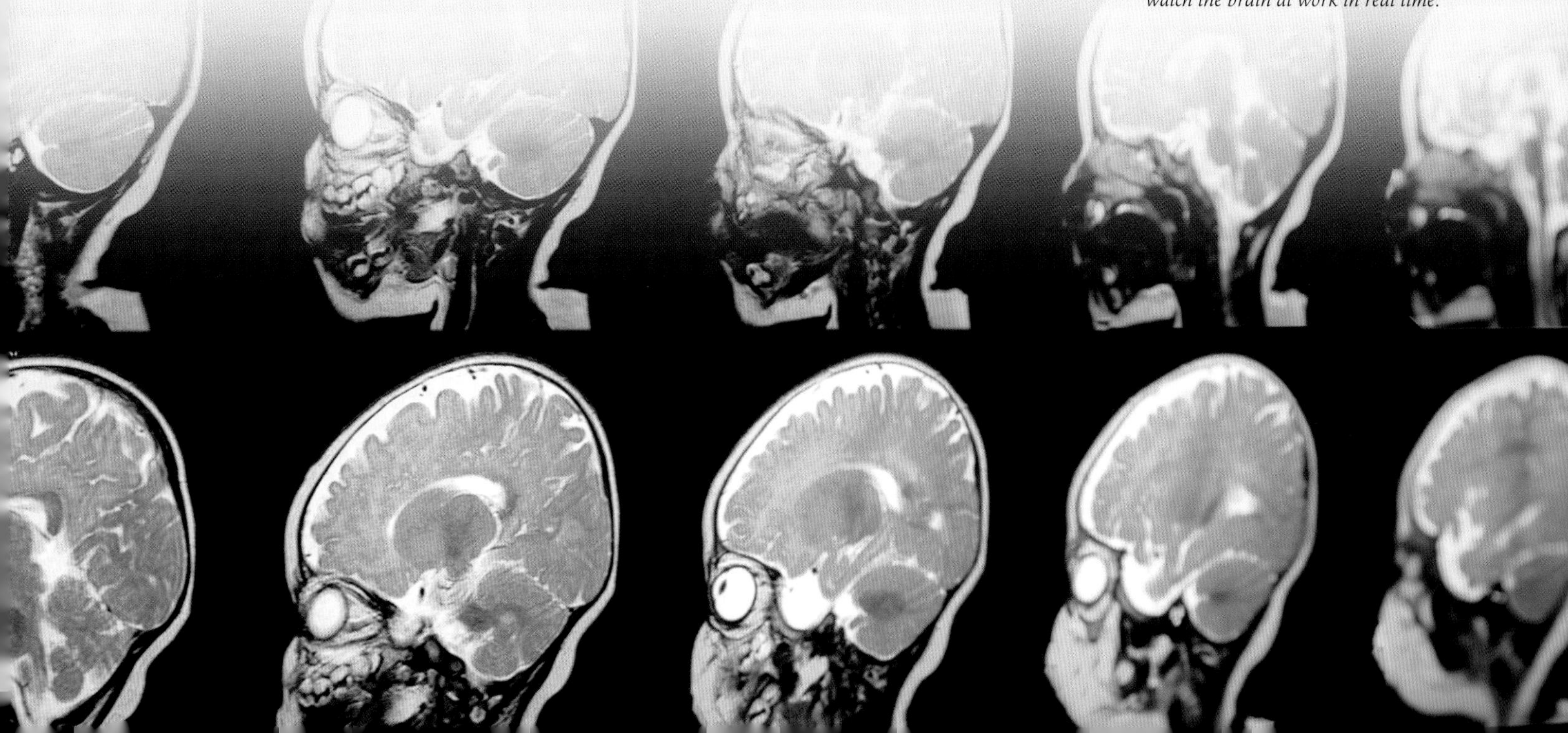

Inside the head

In 2009, the Human Connectome Project was launched: Its goal to build the "wiring diagram" of the human brain. This ambitious project is ongoing and if successful will shine a bright light on how all the parts of the brain work together. And there are many parts. The anatomy of the brain is classified into layers of complexity. For example, the upper level concerns the large structures, such as the cerebellum and the cerebral hemispheres, which early anatomists identified centuries ago. Closer study reveals increasing subdivisions in structure and in function. At the smallest scale is the engram, a physical record of a memory or a thought made up of brain cells connected together. Or at least we think it is. So far the finest details of how the brain fits together have proved elusive. But the search is well and truly on.

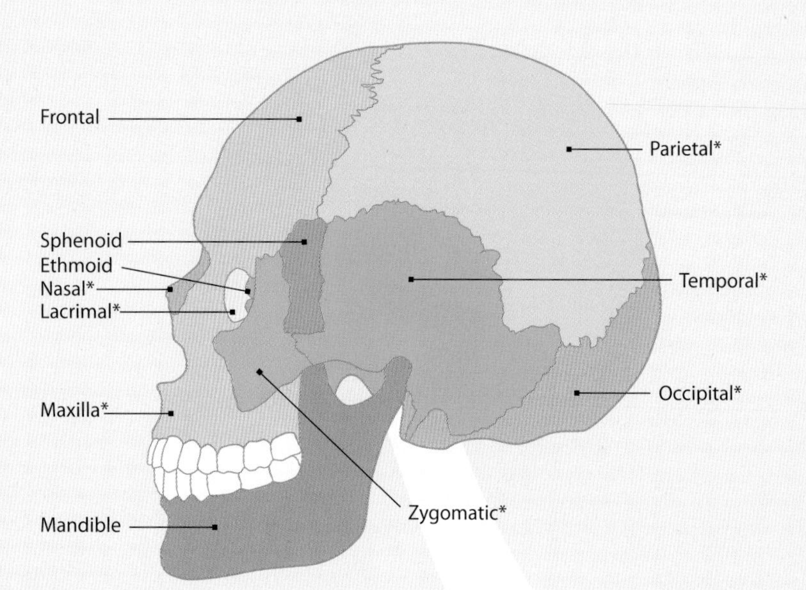

Skull bones

The skull (and jaw) is made up of 22 bones. Some are paired (marked with *), while the vomer and two concha bones in the nose are not shown.

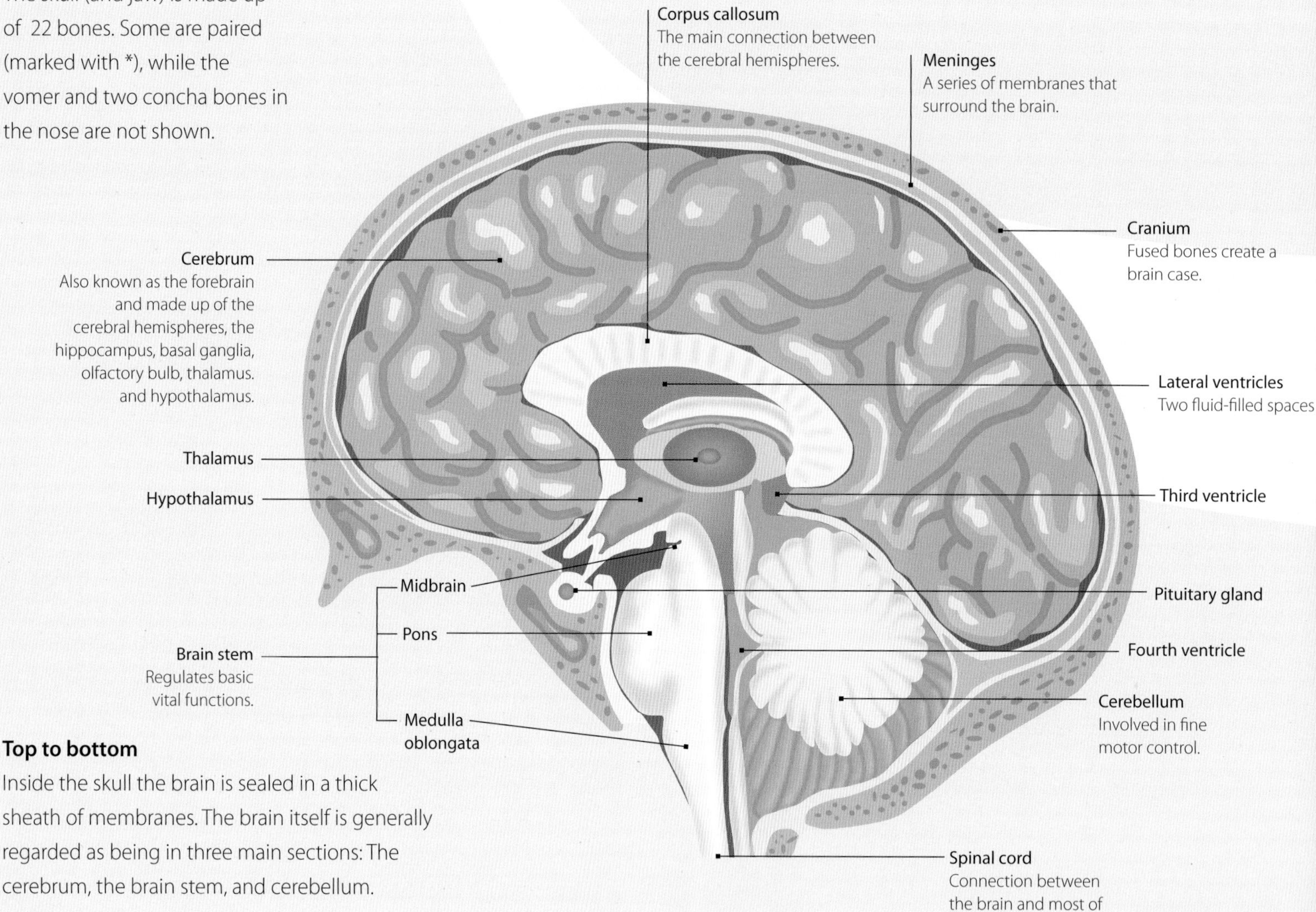

Top to bottom

Inside the skull the brain is sealed in a thick sheath of membranes. The brain itself is generally regarded as being in three main sections: The cerebrum, the brain stem, and cerebellum.

FUNCTIONAL ANATOMY

The brain can also be described in terms of its functions, that is, which bits do what. The localization of functions is still open to debate, with many brain regions seemingly involved in several different things. However, these colorful diagrams shown here give a good idea of what is happening in your head. Diagram 1 is the side view; 2 is an internal cross section; and 3 is a view from above.

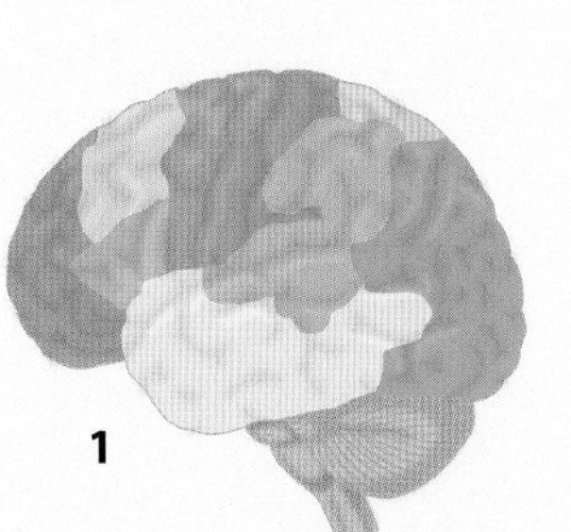

1

- High mental functions
- Eye movement
- Speech
- Voluntary motor function
- Association
- Sensory
- Hearing
- Language comprehension
- Vision
- Coordination

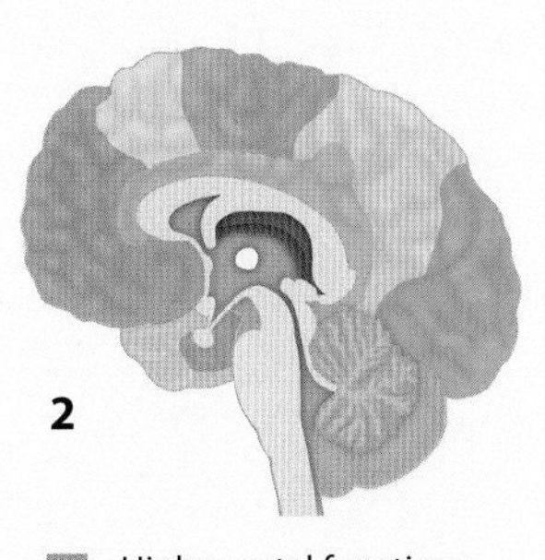

2

- High mental functions
- Eye movement
- Emotion
- Sensory
- Somatosensory association
- Vision
- Voluntary motor function
- Coordination

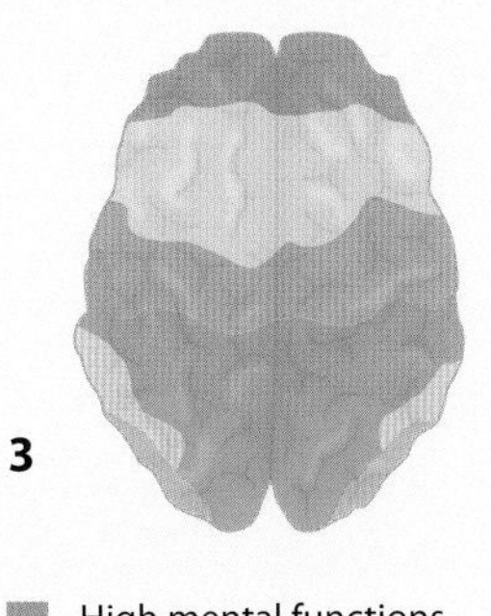

3

- High mental functions
- Eye movement
- Voluntary motor function
- Sensory
- Somatosensory association
- Language comprehension
- Vision

Precentral gyrus
The main movement control center

Postcentral gyrus
The region processing the sense of touch

Frontal lobe
Associated with planning, motivation, and attention

Parietal lobe
Integrates sensory information with knowledge.

Lateral sulcus
A prominent fold, also named the Sylvian fissure for the 17th-century doctor Franciscus Sylvius.

Occipital lobe
Concerned with vision

Temporal lobe
Associated with language, memory, and emotions.

Brain stem

Cerebellum

Spinal cord

Cerebral lobes

The cerebral hemispheres make up most of the forebrain. There are two hemispheres, each one divided into four lobes.

1 A Hole in the Head

THE BRAIN IS SPECIAL. EVIDENCE FROM OUR PREHISTORY APPEARS TO SHOW THAT WE HAVE LONG UNDERSTOOD THAT, either instinctively or through experience. One reason we know this is that the remains of ancient human skulls reveal that even prehistoric societies had a go at brain surgery.

HEAD TARGET

Many hominid skulls have been found with blunt trauma injuries—basically, their heads had been bashed in. This is a hint that even our most primitive ape-like ancestors knew that when it came to defeating a rival, it was best to aim for the head.

In the early 19th century, Alexandre François Barbie du Bocage, a French geographer, came into possession of an ancient human skull. The skull, which had been dug up in northern France, was not wholly intact, but had a sizeable hole in it. Earlier gentlemen scientists had found similar specimens—they seemed especially common in France. Du Bocage recognized this for what it was, a deliberate removal of bone, not the product of accidental injury. He was aware that in medieval times—and still on occasion in his own day—doctors performed a procedure where a portion of the skull bone was cut away. The practice is known as "trepanning" (also termed trephinning), which is derived from an old French word for drill or borer, and gives us a terrifying image of how it was carried out.

Du Bocage's skull specimen showed signs that the bone had had a chance to heal, so the injury had not led to the death of its owner. In other words, this person had undergone trepanning. Nevertheless, the true significance of this find was lost on du Bocage, and the rest of the scientific community for several more decades.

Stone Age surgeons

By the turn of the 20th century, a significant haul of trepanned skulls had been accrued across Europe. The startling fact was that these specimens dated back to at least 10,000 years ago, perhaps even further. This put them into an era where the region had no permanent settlements, no agriculture, and no metal-based technologies. In other words they were Stone Age people, who had undergone some Stone Age surgery.

One can only guess, with some

This skull specimen shows three trepanning holes. To have a chance of success, the cutter would have to stop short of puncturing the membrane that surrounds the brain.

horror, at what instruments were used. The sharpest things at the disposal of the "surgeon" would have been volcanic glass and flints knapped into cutting edges, or large seashells. Whether these tools were used to scrape away the bone or chisel out chunks is open to conjecture. Most holes are in the parietal bone, which covers the side of the head and is reasonably easy to get to and remove. In later specimens, where we presume the tools had become more effective, the holes are more commonly made in the frontal bone. Perhaps this forward location was regarded as being more effective.

Trepanning was probably practiced the world over, certainly in Asia and Europe. Another hot spot was Peru, where almost half the occupants of some grave sites survived at least one trepanning.

Remedy or ritual?

We do not know why prehistoric communities cut holes in their heads. A modern form of the procedure, the craniotomy, is used to remove skull bone to relieve pressure acting on the brain. It has been suggested that ancient trepanning was a treatment for skull fractures. However, specimens frequently have more than one hole, which would appear to rule the treatment theory out. The medieval reasoning for trepanning was analogous: It released demons that had infested the head and were causing headaches, fits, and visions. We can imagine Stone Age elders following a similar line of thought when faced with a clan member suffering similar symptoms. The sheer frequency of trepanning in New World communities would suggest that even healthy people underwent the surgery. Perhaps it was believed to open them to supernatural forces. Whatever the reason, we can see that even our most primitive ancestors understood that the head was the key to their well being.

Above, a bronze knife, or tumi, like the ones used in Peruvian trepannings in the first millennium CE.

THE WOUNDED MAN

The Lascaux Caves in southern France contain some of the oldest works of art known. Most of these paintings from 17,300 years ago are of animals. Only one shows a human figure, who appears to be injured, perhaps by the auroch, or ancestral cow, that is charging at him. Beside the so-called Wounded Man is a bird, interpreted by some as the man's soul leaving as he dies. The bird-shape of the man's head suggests that the artist linked the soul with the head, making it the location of a person's life force.

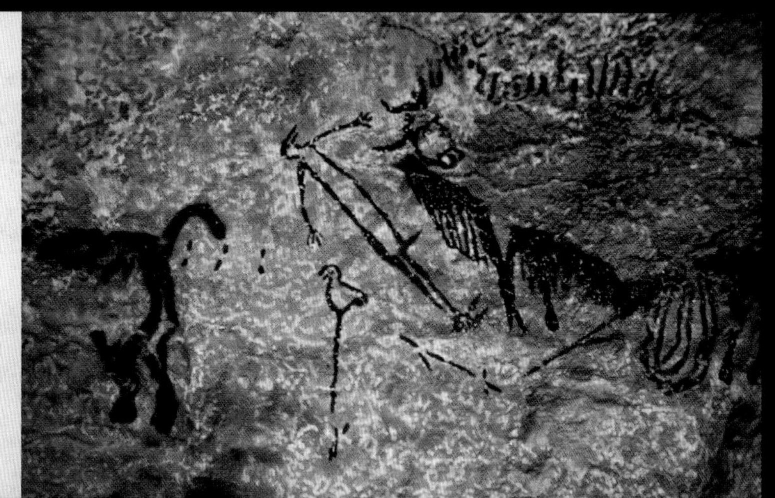

2 The Ancient Egyptian Brain

THE FIRST DOCTORS RECORDED BY HISTORY WERE ANCIENT EGYPTIANS. They were part physician, part priest, and showed a good deal of practical knowledge. However to them, the heart was the primary organ; the brain was rather unimportant.

The earliest examples of medical textbooks were Egyptian papyruses that have survived the intervening centuries. One of them, the Ebers Papyrus, was found tucked between the legs of a mummy dating from 1550 BCE. A second, thought to be a few decades younger, is the Edwin Smith Papyrus (named for the English egyptologist who curated it in the 19th century). It carries the words, rewritten many times, of Imhotep, who lived more than a thousand years before. (As well as being the first named doctor in history, Imhotep was also the architect of the first Egyptian pyramid, which was built at Saqqara for his pharaoh Djoser in about 2600 BCE). According to Egyptian medical doctrine, good health arose from the heart from where channels containing blood, air, mucus, and all manner of other bodily fluids flowed around the body. An injury, to the brain or elsewhere, needed to be treated because it was blocking said channels in some way.

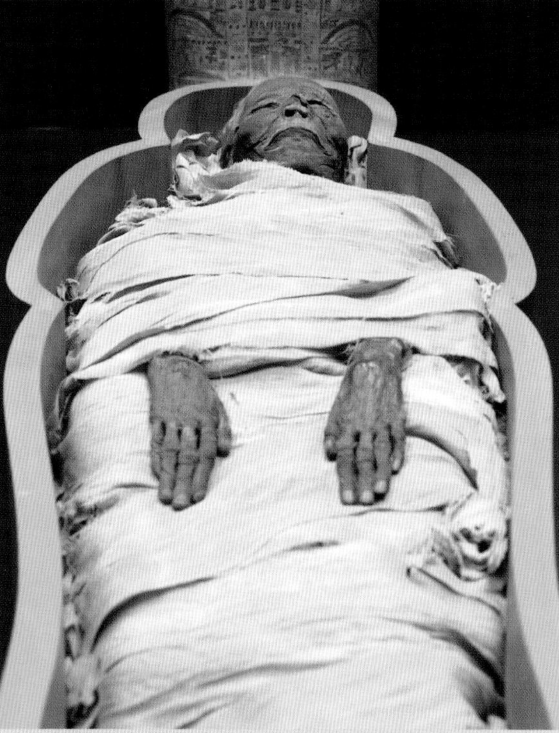

MUMMIFICATION

The ancient Egyptians are famed for their mummifications. These aimed to preserve the body of the dead so it could be reused in the afterlife. The heart was believed to be the seat of emotions and thought and carried a record of a person's deeds in life. The god Anubis would weigh it after death to see how weighed down it was with guilt. Therefore, the heart was left untouched during mummification, while liver, lungs, and stomach were removed and carefully packed into jars. By contrast, the brain was so insignificant it was scraped out through the nose and thrown away!

Mooring up

Imhotep and his later commentators were aware that injury to the head could lead to symptoms in other parts of the body, such as a loss of movement on one side. Their discussions of brain injuries make rather gory reading. After examining a patient to see if the skull is fractured, the advice is for the doctor to put his fingers in the hole. This should elicit the patient to "shudder exceedingly." If other symptoms include swelling, bleeding from the nose and ears, and an inability of the patient to turn his or her neck, then the treatment was to "moor him at his mooring stakes." In other words, the patient was put to bed and nature would take its course, just as a tied-up Nile riverboat rose and fell of its own accord in the waters.

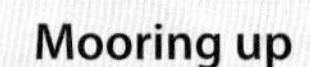

Imhotep, the first doctor in history, was revered as a god by later generations.

3 The Evil Eye

THE EYE IS SAID TO BE WINDOW TO THE SOUL, AND IT WOULD APPEAR that is a sentiment that has held true for millennia. In ancient times diseases of the eye indicated ill health of the mind.

Every ancient culture has its founding medical figure. Egypt has Imhotep, Greece has Hippocrates, while India celebrates Sushruta. All of these figures record many treatments of eye disease, which range from washing eyes with urine and feces to cataract operations not too out of place in modern ophthalmology.

Eye problems in ancient times were often caused by parasites and infections, which are mercifully less common today thanks to modern medicine. However, at the time these diseases, anything from cataracts to a squint, were said to be caused by evil spirits. More importantly those spirits were contagious. If an afflicted person stared at you too intently—and you met their gaze—you too would succumb to the Evil Eye.

4 The Brain in China

TO THIS DAY CHINESE MEDICINE HAS TAKEN A DIFFERENT PATH TO ITS WESTERN COUNTERPART. It elevates the brain to a special status, although it is not seen as one of the major organs.

Chinese medicine is based on the principle of Zang-fu, which divides body parts into two groups. The major organs, the zangs, number five, each one linked to one of the five traditional Chinese elements: Wood, earth, fire, water, and metal. The zangs are the lungs, kidneys, heart, spleen, and liver. The fu organs include the stomach, intestines, and the brain. The brain is a special type of fu, but emotions (and their related mental states) nevertheless arise from the activities of the zang organs. The heart creates happiness (and intelligence), the kidneys make you scared, lungs sad, liver angry, and the spleen controls awareness.

BALANCING OUT

Good health, according to Chinese medical thought, is due to a balance of the basic forces of yin and yang. One of the means to rebalance an ailing body is acupuncture, which uses tiny needles pushed into the skin. It is thought that this practice arose from the belief that the demons believed to cause imbalances could be driven out by puncturing the skin, thus making the body an uncomfortable place for the malevolent spirit.

5 Hippocrates and his Humors

NEUROSCIENCE BEGINS WITH WESTERN MEDICINE, AND THAT BEGINS WITH HIPPOCRATES. Nevertheless, even this medical pioneer employed a lot of mystical ideas in his theories of disease.

THE FURIES

According to Greek mythology, a person could be driven to hatred, vengeance, and eventual madness by the arrival of underworld deities called Erinyes, or more commonly the Furies. Once you had attracted their attention, these subterranean crones would hound you relentlessly until you did their dreadful bidding. Writers would refer to them with flattering euphemisms., such as the "Kindly Ones." A direct mention of them was too risky.

Hippocrates lived on the island of Kos in the 4th century BCE, during Greece's Golden Age. In keeping with many of his fellow Greeks, Hippocrates refused to accept that disease, including nervous disorders like paralysis and fits, were caused by demons that had infested the body—through the eyes or any other portal. Instead, he looked for a physical reason for ill health, and it is thanks to Hippocrates that doctors to this day diagnose and treat diseases according to the symptoms patients present.

According to Hippocrates's humoral theory, the physical characteristics of the four elements were closely allied with the medical and emotional impact within the body.

Yellow Bile
hot
dry
FIRE
AIR
EARTH
WATER
Blood
Bile
wet
cold
Phlegm

From elements

At the time Greek scholars believed that the world—including the human body—was made up of basic substances, later known as the elements. That kind of idea was reflected in Chinese and Indian theories, but in Greece there were just four elements: earth, air, fire, and water.

These elements existed in the body as liquid humors: blood, phlegm, yellow bile, and black bile. Each shared the characteristics of their element and impacted emotions. Air-filled blood created a sanguine optimism. Watery phlegm imparted a phlegmatic calm. Too much earthy black bile weighed people down with melancholy, while yellow bile fueled a fiery rage.

A head injury required that the humors that gathered there as pus be released. If the skull was fractured all to the good. If not, then Hippocrates would drill an opening. Despite its roots in an untested theory based on false assertions, this craniotomy procedure was frequently a sensible treatment that relieved swelling.

6 Theories of Vision

ANCIENT GREEK PHYSICIANS MADE DETAILED EXAMINATIONS OF THE HUMAN EYE, AND NOTED MANY NEW FEATURES, INCLUDING THE VITREOUS HUMOR AND OPTIC NERVE. However, the mechanism by which the eye worked was beyond their understanding. Nevertheless, they had a few goes at explaining how the sense of vision operated.

In the days of Hippocrates, the dissection of dead human specimens was not allowed, and even cutting up animals to learn more about anatomy was seen as morally dubious. However, by the following century, the new Greek city of Alexandria at the mouth of the Nile became a haven for freethinking scholars and researchers. The rulers of the city had ambitions to turn Alexandria into the knowledge hub of the known world, and so allowed human dissection as a necessary evil. Herophilus became the top dissector in the city, and is regarded as a founder of the science of anatomy. He worked on the corpses of executed criminals—and may even have cut up some people who were condemned to die and then did so on his dissection table. Herophilus found that the eye had a transparent core and was connected to the brain by a thick nerve.

"The eye obviously has fire within, for when one is struck, this fire flashes out."

THEOPHRASTUS, 3RD CENTURY BCE

Inside and out

This new understanding of the eye led to two competing theories about how the eye worked. Alcmaeon was the leading proponent of what became known as the "extramission theory." This held that the eye's watery center was able to emit a flash of "fire"—not necessarily visible, but more akin to heat—which traveled out of the eye and reflected off objects. The eye was therefore like a flashlight, sending out a beam that illuminated whatever was before it. Alcmaeon's proof was that an impact on the eye created a blinding flash as the mechanism was disrupted. Vision became difficult in the dark because the emitted fire was blocked by its opposite entity, darkness.

Later thinkers, including Epicurus and Aristotle, preferred another theory, known as "intromission." This held that objects gave out an invisible imprint of particles, which traveled to the eye and then fired up the internal light in the eye, creating the vision.

An illustration in Johann Zahn's 17th-century book Oculus Artificialis (The Artificial Eye) *compares the extramission theory of vision (top) with a more modern optical interpretation of how light and the eye behave.*

7 Three Souls

THE ANCIENT GREEKS ALSO DEVELOPED IDEAS ABOUT HOW A PERSON PERCEIVED THE WORLD AND UNDERSTOOD IT. These theories were centered around the soul, but arguments raged about its physical nature and where it was located in the body—the heart, the head, or elsewhere.

"Human behavior flows from three main sources: Desire, emotion, and knowledge."

PLATO

We all understand the idea of taking something important "to heart" and being "heartbroken" when things go badly wrong in life. This concept stems from the ancient notion that at least part of our consciousness is harbored by the heart. One dominant ancient Greek idea was that the human body was governed by three souls: The rational soul that did all the thinking was located in the head. The heart controlled our passions—as we still characterize it today—while the basic driving appetites that maintain the body were governed by the third soul in the liver.

One of the first proponents of the tripartite soul was Democritus, who is better remembered for proposing that nature was composed of tiny units termed "atoms." An object's characteristics were defined by its

LIFE IN THE CAVE

Plato's most celebrated contribution to philosophy is the Allegory of the Cave. This is a metaphorical account of how he understood a person's ability to perceive the world. At birth, humans found themselves as if facing the wall of a dark cave. True reality lay outside in the sunlight, and a person's perceptions were merely shadows that were cast on the wall by real things. Reality was constructed of what Plato called "eidos." We get the word "idea" from this, but we translate it as Form. All things had an eternal Form, said Plato, and we have to use the intellect to escape the illusory world of our perceptions.

Plato's theory of the soul was that all knowledge was contained within a person's intellectual soul. It was up to its owner to reveal this knowledge through a philosophical life.

atoms: Fire atoms were prickly while water atoms were smooth and slippery. By the same token, the atoms of the soul were tiny and imperceptible as they gathered in the organs. Democritus believed that the soul's atoms dissipated on death, like the rest of the body, but Plato suggested that the intellectual soul was immortal and passed from body to body. His idea was that different members of society were controlled by different souls, mirroring the social hierarchy. Philosophers like him were commanded by their intellect. Military types were ruled by their hearts, while the lowly peasants answered only to their livers. Plato's star pupil, Aristotle, was less convinced. To him the head was merely a radiator for the heart, where the excess heat generated by human passion was dissipated into the air.

8 Ancient Theories of Sleep

According to Homer, the great figure of Greek literature, sleep and death were "two twins of a winged race." In other words being asleep brought you closer to death.

The rejuvenating powers of sleep were well understood even in premodern times, yet the hours of slumber were regarded with a mix of trepidation and mystery.

The inextricable link between sleep and darkness led ancient thinkers to consider it a murky, almost evil force, where a person drifted on the edge of the underworld. However, others attempted to describe it in more earthly terms. Empedocles, credited with formalizing the four-element approach to nature, suggested that people fell asleep because the blood grew cold as its fire withdrew at night. Alcmaeon proposed that during sleep blood drained from the brain. If it did not return, then sleep would convert to death.

Time to digest

Aristotle had other ideas. He thought sleep was the result of eating. Fumes from food entered the blood and floated up into the brain, which was tasked with releasing heat. The fumes accumulated in the brain, and as it cooled at night, the fumes sank back into the body, where they cooled the heart, reducing its ability to function. One of its functions was to create awareness—and as this reduced, sleep followed.

THE SONS OF NYX

In the pantheon of Greek gods, the night was ruled by Nyx. Nyx was the daughter of Chaos and patrolled the night dressed in a starry tunic aboard a horse-drawn chariot. Nyx had two sons with Erebus, the god of darkness. The first was Thanatos, the god of death, who tortured sinful souls in the afterlife. The other son was Hypnos, a more gentle figure who ruled sleep.

9 Galen's Pathways

THE MOST INFLUENTIAL FORCE IN NEUROSCIENCE FROM THE ANCIENT WORLD WAS GALEN, A GREEK DOCTOR. HIS WORK RELOCATED CONSCIOUSNESS IN THE HEAD—and it has stayed there ever since. Galen's work on the brain's anatomy was instrumental in mapping the nervous system.

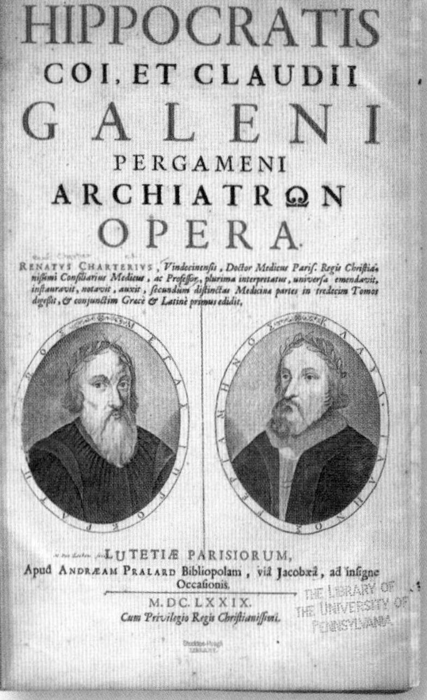
HIPPOCRATIS
COI, ET CLAUDII
GALENI
PERGAMENI
ARCHIATRΩN
OPERA.
RENATVS CHARTERIVS, *Vindocinensis, Doctor Medicus Paris. Regis Christianissimi Consiliarius Medicus, ac Professor, plurima interpretatus, universa emendavit, instauravit, notavit, auxit, secundum distinctas Medicinæ partes in tredecim Tomos digessit, & conjunctim Græcè & Latinè primus edidit,*
LUTETIÆ PARISIORUM,
Apud ANDRÆAM PRALARD Bibliopolam, viâ Jacobæâ, ad insigne Occasionis.
M.DC.LXXIX.
Cum Privilegio Regis Christianissimi.

The Galenic corpus, *the combined writings of Galen, became the leading medical textbook for the next 1,300 years.*

By the 2nd century CE, the epicenter of knowledge had shifted from the Greek cities in the Eastern Mediterranean to the Roman ones in the West. Medics from Greece flocked to Rome to ply their trade. One of them was Galen, a doctor who had been trained at Pergamon's Asclepeion, or healing temple, the most renowned medical school in the ancient world.

"The best physician is also a philosopher."

GALEN

Monkey brains and more

Galen was a devotee of Hippocrates, and a fan of Aristotle's empirical methods in that he based his medical ideas on what he witnessed. However, what he witnessed put him at odds with the basis of Aristotle's medical theories. Galen performed much of his anatomical research on monkey specimens, which he found to resemble the human body a lot more closely than his other subjects which included pigs, sheep, weasels, and an elephant. Human dissection was outlawed, but Galen had other means of getting a glimpse inside the human body (*see* box).

In the year 177, Galen gave a lecture titled *On the Brain*. In it he rubbished the idea that the brain was some kind of heat regulator. If that were the case, he reasoned it would have been placed nearer the heart—and the senses would not be attached to it. Galen's great contribution was to show that a series of nerves connected the brain to the body. Some of the nerves were pathways from the senses, while others went to the muscles.

He demonstrated this by carefully slitting a pig's throat, severing the nerve that connected to the larynx muscles. The pig was able to breathe and its heart continued to beat. However, it was unable to squeal. This nerve is now known as the recurrent laryngeal nerve, but is also remembered as Galen's nerve for its discoverer.

GLADIATOR'S DOCTOR

Galen's medical techniques drew him into frequent conflict with the Roman physicians, who were mostly quacks. Nevertheless, Galen caught the attention of the imperial classes, and served as court physician to Emperor Commodus (the bad guy in the *Gladiator* movie of 2000). Galen had earlier served as the camp doctor for a team of gladiators. He was very good at it: It is reported that only five gladiators died on his watch. The posting also gave Galen a good look inside the living human body.

10 The Brain in Section

GALEN'S TEACHINGS FORMED THE BASIS OF NEUROSCIENCE FOR MORE THAN A THOUSAND YEARS. Those who came after sought to figure out what each bit of the brain did.

The brain worked using "animal spirits," according to Galen. These mysterious fumes or fluids were distilled from the more basic life-giving vital spirits that traveled up from the heart to the brain through the carotid artery. (The waste residue of this process was phlegm, which drained out through the nose.) The spirits were stored in the ventricles of the brain—four seemingly hollow spaces located deep inside. When needed, the animal spirit would be tapped from these reservoirs and then directed along nerve pathways to gather sensory information or command muscles into action.

A diagram from the 16th century by scholar Albertus Magnus imagines what the ventricles looked like inside the brain. Christian images preferred to show three ventricles, not four, to match up with the Holy Trinity.

Localizing functions

In the 4th century CE, Nemesius, the Bishop of Homs in modern-day Syria, tried to add more detail and splice in a bit of Christian dogma. He proposed that the ventricles filled with a spirit (holy or otherwise) were responsible for different functions. The two forward ventricles, which protrude into the front of the brain, were used for perception. Thoughts were generated in the central chamber (known as the third ventricle), while the rearmost, fourth ventricle housed the memory. This was in keeping with Galen's observations on function of the brain. Sensory nerves appeared to be mostly linked to the front of the brain, so it made sense that the perception center was there.

Motion and memory

The back of the brain was more associated with the motor nerves, which linked to the muscles. St. Augustine of Hippo, a theologian from Algeria, weighed in with the idea that the rear ventricle controlled motion, and he merged memory with understanding in the middle ventricle. Those on either side of the argument pointed to evidence from injuries, where a blow to the front or back of the head affected whichever bodily function suited their case. However, the next contribution would come when someone wondered what the brain would do if it were disconnected from the body.

11 Flying Man

IBN SINA, BETTER KNOWN IN EUROPE AS AVICENNA, WAS A SCHOLAR DURING THE GOLDEN AGE OF ISLAM. His interests lay in the nature of the soul, and how it interacted with the body. To find out more he used a famous thought experiment.

Avicenna was Persian, but like most 10th-century scholars, his investigations began with the study of the works of Greek philosophers.

When it came to consciousness, Avicenna sided with Plato, who said (unlike Aristotle) that the soul survived death. That meant the soul—we might characterize it as the mind, today—was separate from the body. And to prove it, Avicenna imagined the Flying Man. The Flying Man was blindfolded, ears plugged, and suspended by some force or other in midair. His arms and legs were held away so he could not touch any part of his body. He spent his entire life like this, devoid of sensation. Avicenna felt sure that if he were the Flying Man, he would have had a sense of himself, that self exists separately from the body. It does not need to know it even has a body associated with it. Avicenna's concept of a mind separated from the body was an early rendition of "dualism," and this idea would be central to later investigations into how the brain worked.

12 Optical Eyes

THE FIRST TRULY SCIENTIFIC BREAKTHROUGH IN NEUROSCIENCE INVOLVES THE FUNCTIONING OF THE EYE. It was made by the prolific Arab scholar Ibn al-Haytham who pondered the behavior of light beams while languishing, so the story goes, under house arrest.

They say "Seeing is believing," and it was al-Haytham who showed we can trust our eyes. Al-Haytham was known in medieval Europe as Alhazen (or Alhacen) hailed from Basra in Iraq. He made a name for himself as a mathematician, astronomer, and engineer, and began to court the Fatamid caliph, who lived in Cairo, now the capital of Egypt. Alhazen proposed a plan to dam the Nile at Aswan and so give the caliph control over Egypt's most valuable resource. The proposal turned out to be a bit rash. In 1011, Alhazen was summoned to Cairo, where he soon found that damming the mighty river was beyond even his abilities. He weighed up his options and opted for feigning insanity as the best way to avoid the caliph's wrath. The caliph responded by placing Alhazen under house arrest, where he would remain for the next decade.

SCIENTIFIC METHOD

Alhazen is one of the first historical figures to design fully scientific experiments to test his ideas. He also used mathematics to predict the outcomes of experiments. It would be another 600 years or so before this way of doing science was properly formalized.

Book of Optics

It was while locked away in Egypt that Alhazen performed the work for which he is most remembered. The fruits of his research was *Kitab al-Manazir (Book of Optics)*. In it Alhazen relates how he used one of the first properly scientific experiments to prove that beams of light travel in straight lines. He observed a distant candle light through a hollow tube. He then blocked the far end, and the flicker of light was no longer visible. Obvious it may be, but this was the first empirical proof that light could take only one straight path to the eye.

This is the first evidence that disproved earlier theories of vision. The eye did not send out light or fire to illuminate objects, nor did objects transmit some kind of impression. Instead, objects radiated light beams in all directions, the paths of which obeyed simple geometric laws—and a few of them traveled into the eye. Alhazen realized that the eye behaved in the same way as a *camera obscura* (meaning a "dark room"), where light enters through a single hole in the wall. That narrow beam of light projects an inverted image on the far wall. This is repeated in the eye, with light entering the pupil and forming an upside down image on the retina. At that point guesswork kicked in again, as Alhazen fell back on the concept of a spirit that transfers the image to the brain. Nevertheless, progress was being made.

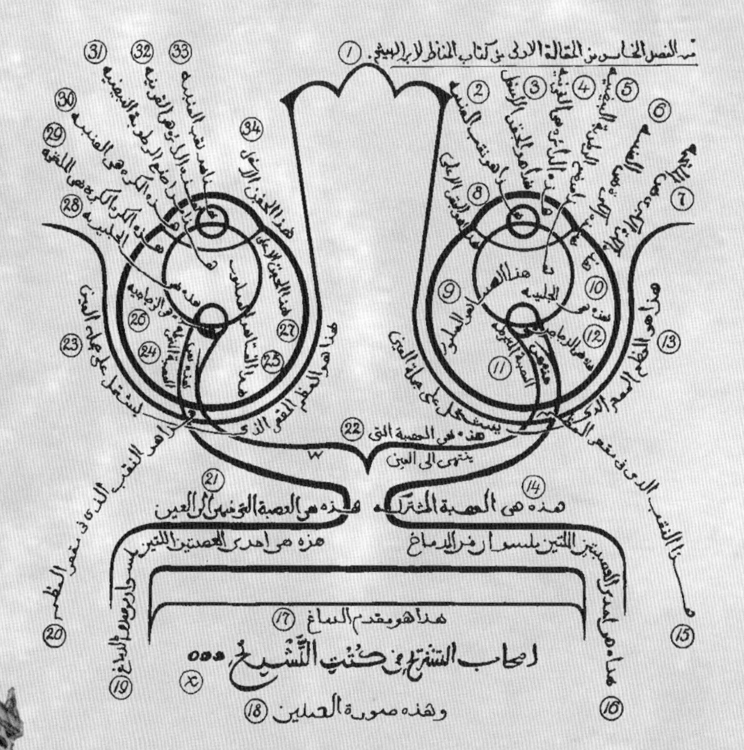

To Alhazen, a camera obscura and the eye created internal images in the same way. His diagram of the eye (above, the numbers have been added later) showed a good understanding of the anatomy of the optic nerves, but still relied on the idea that the nerve was some kind of pipe for a spirit that conveyed the image.

13 Passions and Emotions

DURING THE MIDDLE AGES, THE DEBATE AROUND EMOTIONS AND PERSONALITY MIRRORED THE IDEAS PUT FORWARD IN CLASSICAL GREECE. Humans differed from the animals in that their heads held sway over the passions of the heart and brutish desires of the liver.

Thomas Aquinas

Galen, who dominated all thoughts medical until the Renaissance, described personalities in terms of a battle between the three seats of the soul: The liver, heart, and brain. Only humans were able to use their intellect to control the animal drives that took hold of the lower animals. How well each person did this depended on the proportions of Hippocrates's humors in their body. Thomas Aquinas, a 13th-century Italian monk, made a life's work of fusing Greek science with Christian dogma. Christian teaching relies on emotive terms to describe the deeds and motivations of key figures. Aquinas introduced the idea that emotions and intellect were not entirely separate, and could modify each other. For example, he identified a hierarchy of three types of love: Amorous desire, respect, and charity.

14 Dancing Mania

IN THE 14TH CENTURY, EUROPE WENT DANCE CRAZY. Entire villages suffered episodes of uncontrolled jerky spasms, as if dancing to some unheard rhythm. Something had taken control of their muscles and minds.

From the 1370s, incidents of manic dancing were reported across northern Europe from England to Switzerland. The episodes were characterized as "choreomanias," meaning "dance madness." The Greek word for dance—*chorea*—is still used to describe the involuntary movements of certain congenital nervous disorders. However, the dancing mania of Europe appeared to be communicable, spreading as an epidemic across the land.

SALEM WITCH TRIALS

As with the dancing mania, ergot poisoning is also suspected as being behind many of the witch hunts of 17th-century America—most notably at Salem, Massachusetts. In 1693, 20 people were put to death accused of practicing black magic. Reports of spasms, visions, and delirium among the townsfolk were seen as demonic possession, but were probably the result of fungus in the previous year's rye harvest.

The horn-shaped ergot fungus attacks an ear of rye.

Fire and music

The accounts of the dancing are confused and contradictory. Many occurred during religious congregations, which hints as least some of the dancers were caught up by mass hysteria. However, there is a recurring theme of delirious visions and frenzied movements. Musicians played along with the dancers in the hope that it would help to calm them. Generally, attacks continued until people collapsed with exhaustion. In Taranto, Italy, the dancing was blamed (erroneously) on spider bites, and a traditional dance—the tarantella—lives on as a folk memory of the mania. (The tarantulas of America are named for these spiders.) Some dancers also had extreme burning pain in their hands and feet—named St. Anthony's Fire. This symptom points to the cause of the mania as ergot poisoning. Ergot is a fungus that grows on grain stored in damp conditions. The fungus has a powerful effect on the nervous system, leading to paralysis, convulsions, hallucinations, and insomnia. The "fires" that attack the extremities are due to the blood being restricted to the hands and feet.

15 Da Vinci's Waxworks

LEONARDO DA VINCI IS HAILED FIRST FOR HIS ART, AND AFTER THAT FOR HIS FUTURISTIC ENGINEERING DESIGNS. In 1506, he also began to record anatomical features, but found nature was against him.

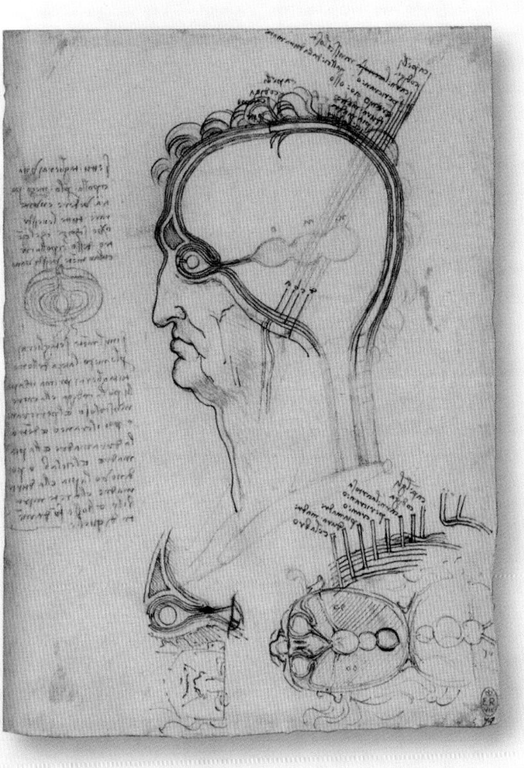

Many of da Vinci's anatomical drawings date from around 1510. They were made with the help of Marcantonio della Torre, the professor of anatomy in Padua.

Leonardo's great fame stems from his ability to depict the human form. He had been a student of anatomy since childhood, and once established as a great painter he was given access to corpses at Florence's hospital. He had to work fast to record structures before they dried out or putrefied.

The contents of the skull proved a problem. Leonardo was intrigued because he had experimented with live frogs to show that life continued if the heart was removed (for a little while at least), but ended abruptly when the brain was severed. Leonardo believed the life-giving soul lay in the ventricles, and to get a better look, he bored a hole in the back of the head and pumped in hot wax—forcing any residual liquid out of a second hole on top. The ventricles are interconnected, and so once set and the skull and brain removed, the wax created a three-dimensional representation, but of what? A man's imagination, memory, even his soul?

16 Michelangelo's Hidden Brain

LEONARDO DA VINCI WAS NOT THE ONLY RENAISSANCE ARTIST TO SHOW AN INTEREST IN THE HUMAN BRAIN. Michelangelo also had a secret fascination with it, and he left clues to his in-depth knowledge of its anatomy in his most famous masterwork: the frescos of the Sistine Chapel.

Michelangelo spent four years painting the ceiling of the Sistine Chapel in the Vatican. It contains many of the best loved examples of Renaissance art. Michelangelo took his time, and there are no mistakes, not accidental ones at any rate. The artist himself appears twice, on some flayed skin and as a severed head—some say this shows that he did not like the Medici family who'd commissioned the work. The most famous scene from the frescos is the Creation of Adam, where God and the first human touch fingers. God and a host of angels are surrounded by a flowing red robe that looks like the brain in section. A green scarf flutters where the vertebral artery would be, a cherub's arm represents the optic nerve, his leg the pituitary. The legs of the angels form the spinal cord. Michelangelo did not explain his composition but commentators suggest it was his way of opposing the Christian belief that the heart ruled the body. Instead, perhaps he thought the primary gift from God was one of intellect housed in the brain.

Michelangelo appears to be suggesting that Adam's creator was a big brain.

17 Vesalius's Dissections

SCIENCE GOT ITS FIRST PROPER LOOK AT THE BRAIN through the drawings of Andreas Vesalius. His investigation of the physical nature of the brain put an end to any spiritual dimension in neuroscience.

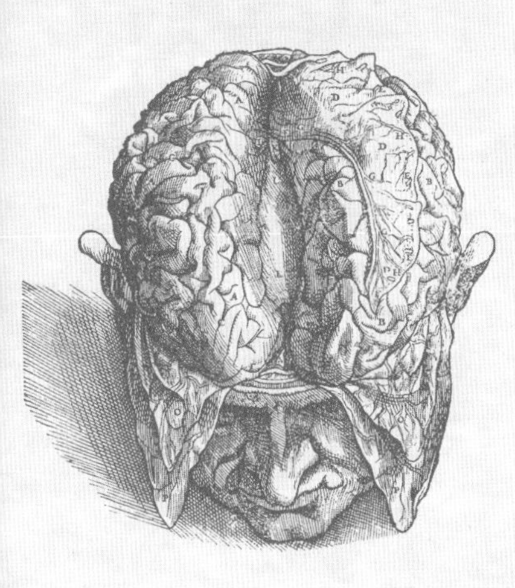

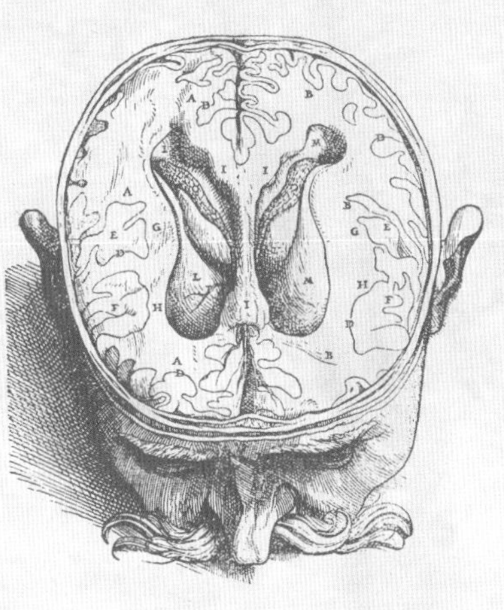

Book 7 of De humani corporis fabrica *covers the brain, from the hair, skin, and skull to the ventricles.*

In the 14th century, the Black Death had swept westward through Asia and Europe, killing a fifth of the world's population. A thousand-year prohibition on the dissection of human bodies was lifted so doctors could attempt to find out more about the disease. They found little to help, but the door was opened to a thorough investigation of human anatomy—and the brain. The first drawing of the human brain dates from 1316, but modern neuroscience is most in debt to the later work of Andreas Vesalius.

Andreas Vesalius is the founding figure of human anatomy. The day he graduated as a medical doctor he was offered the chair of Surgery and Anatomy at Padua's university.

Correcting errors

Belgian by birth, Vesalius worked in Padua, Italy, in the mid-1500s. His seminal 1543 work, *De humani corporis fabrica (On the Fabric of the Human Body),* corrected many misconceptions, long-held since the days of Galen. Among his most important discoveries was that the human brain was not covered by a *rete mirabalis*, or network of blood vessels. Galen had asserted that this network was the structure that distilled the animal spirits used by the brain from life-giving spirits welling up from the heart. A *rete mirabalis* around the brain is found only in hoofed animals, and is indeed a rather marvelous system for cooling their heads as they sprint from danger. In addition, the Vesalius showed that all nerves originated in the brain and not —as had been believed for nearly two millennia—the heart.

Ending myths

While the human brain differed from those of many animals, the structure of the ventricles appeared remarkably similar in pigs, horses, and humans. If a human's ventricles harbored intellect, Vesalius wondered why they did not in dumb animals. Vesalius's work showed that the brain was the primary organ of the body—the seat of the self. However, he had no idea how it worked!

18 A Witches' Disease

IN THE PREMODERN ERA, MOVEMENT DISORDERS, OR CHOREA, WERE THE MOST OBVIOUS FORMS OF NEUROLOGICAL DISEASE. However, such symptoms were frequently seen as a sign of witchcraft!

Many people burned as witches in the Middle Ages are believed to have been Huntington's sufferers.

Huntington's disease is a degenerative disease that attacks the nervous system. Its most obvious symptoms are writhing, rigidity, and odd body posture. In medieval Europe these were regarded as evidence of demonic possession, especially in concert with the marked personality changes that often herald the onset of the disease. Sufferers were at best shunned and at worst put to death as witches. The condition was recognized as a disease by George Huntington, a doctor who lived in East Hampton on Long Island. This community was founded by English settlers in the 1640s, and it became noted for the prevalence of "chorea"—known locally as magrums—among its inhabitants. Huntington's grandfather and father practiced medicine in the town before him, and by the 1870s, George had almost 75 years worth of data about the incidence of the disease. It became apparent that the disease was inherited, and it was in fact the first disease discovered to follow the new science of genetics, itself still in its infancy. However, Huntington was able to offer no cure—and there is none today.

19 Apoplexy

THE WORD APOPLEXY WAS THE GREEK TERM FOR WHAT WE NOW CALL STROKE. It is the oldest distinct neurological disorder, with records of it dating back more than 3,000 years.

Apoplexy means "to be struck away with force," and the modern term, stroke, dates to the end of the 16th century, and is more or less an attempt at a translation. Today, the word "stroke" is used to refer to a problem in the brain, while the meaning of "apoplexy" has been broadened to describe a hemorrhage inside a gland or organ.

Over the centuries, many physicians have attempted to understand stroke. Hippocrates described it this way: "A healthy subject is taken with a sudden pain; he immediately loses his speech and rattles in his throat. His mouth gapes and if one calls him or stirs him he only groans but understands nothing. He urinates copiously without being aware of it. If fever does not supervene, he succumbs in seven days, but if it does he usually recovers."

Hippocrates thought stroke was caused by an excess of black bile that had grown too cold. Hence the need for a fever, which would warm up the bile and promote recovery. Galen prescribed drawing blood using leeches, but admitted it was largely down to the position of the Moon and planets as to whether the patient would pull through or not.

Johann Webfer had a varied career serving as physician to various noble families and was an expert on poisons—although the two were probably unconnected!

Islamic contribution

By the 10th century CE, Baghdad had become the world center of medicine. The director of its hospital was al-Razi, also known as Rhazes. Rhazes described the symptoms of apoplexy as: Collapsing to the ground, snoring without sleeping normally, and not feeling it when pricked with a needle. Rhazes found apoplexy could impair speech and paralyze one side of the body. His conclusion was that the main problem was black bile clogging up the ventricles, and prescribed heating the head with a hot metal rod!

Circulatory problem

The true causes of apoplexy were revealed by Johann Jakob Wepfer, a Swiss pathologist, in 1658. Wepfer had an interest in the blood vessels that supplied the brain, and mapped out the carotid and vertebral arteries that fed blood up through the neck and spinal column. In his book, *Apoplexia*, Wepfer showed that one type of apoplexy could be caused by bleeding inside the brain. This is known as "hemorrhagic stroke." Another had the opposite cause when a blockage to one of the vessels in the brain starved part of it of blood. Today, this is known as "ischemic stroke." Later research found it was blood clots forming in vessels that caused these blockages.

20 Descartes: Reflex and Reason

RENÉ DESCARTES IS REMEMBERED FOR MANY THINGS, SUCH AS THE INVENTION OF ALGEBRAIC GEOMETRY—GRAPHS TO YOU AND ME—PLUS THE FAMOUS PHRASE: *Cogito ergo sum*, "I think, therefore I am." He also made contributions to neuroscience, running from the absurd to the prophetic.

Descartes' interest in the brain stems from his enquiry into the nature of consciousness and awareness. He had been a sickly child and was not a strong man, and had a habit of spending a long time in bed. One morning he awoke only to it again—his first awakening had been a dream. He began to wonder how one could differentiate wakefulness from dreaming. He could not discount the idea that his life so far had been a dream. This ever presence of doubt

To Descartes our senses, such as vision, resulted in perception due to some mysterious interaction with the pineal body (marked as H).

formed the basis of his thinking. He came to doubt every perception; everything he saw, tasted, and touched was the product of a sensory process that he did not understand, and had to doubt. He was forced to admit that it was possible that he, Descartes, was being controlled by some demonic force, which offered a false view of the real world. Unlikely as that was, Descartes could not disprove it, but the very fact that he was doubting his own self, meant that at least his mind (if not his body) must exist. While the hypothetical demon could alter reality, it could not make a non-existent entity doubt its own thoughts. Only something that exists can think, and only thinking things can doubt themselves: *Cogito, ergo sum*.

DUALITY

Descartes' opinion of consciousness forms the foundations of a concept known as "dualism" (although Avicenna had already pondered similar ideas). Dualism insists that the mind—the thinking self—is separate from the body, which is a robot-like machine, or automaton. In the diagram below, published in 1662 after Descartes had died, he shows how stimuli from the eyes or fingertips travel along the nerves to the spirit-filled ventricles. The responses to those stimuli are then mediated by the motions of the pineal body (marked as H).

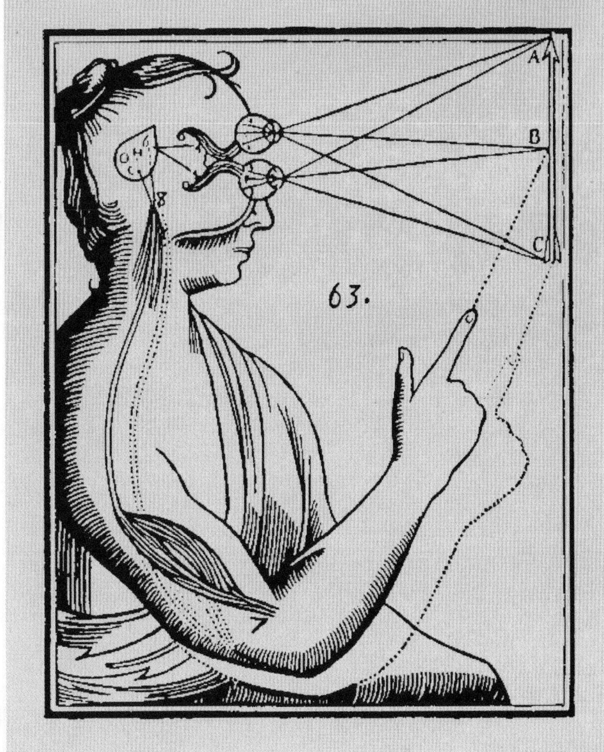

Reflexive system

Descartes had some thoughts on how the brain worked, but struck them from his writings while alive so as to avoid the wrath of Europe's religious authorities. After his death, a work titled *De Homine* (On Man) was published in 1662. In it Descartes describes his theory that the brain and the body worked largely on automatic. The nerves had valves that controlled the ebb and flow of animal spirit through them. As a fingertip deformed when pressed on an object, the valves in the nerve under the skin were opened. Animal spirits flooded into the nerve from the brain's ventricles and caused the arm muscles to move in response to the touch stimulus.

In some instances thought was required to intervene in the process, and Descartes proposed this was done using the pineal body, or gland. The pineal body stands alone (unlike most of the brain's structures, which are paired). It was not part of the brain itself but was nevertheless bathed in cerebrospinal fluids—or "animal spirits" as Descartes understood them. He believed that the flow of animal spirits around the body was ultimately at the control of the pineal body which made minute movements that would modify automatic, reflexive activities. Descartes' theories were not widely adopted. Versalius's argument about the ventricles was even more persuasive here: Many animals, even the most predictable in behavior, had very big pineal bodies, so it was hard to believe that this body held intellect and reason. Nevertheless, Descartes' way of breaking down the problem of the mind and body into first principles would be of help to future neuroscientists.

21 The Circle of Willis

TWO YEARS AFTER DESCARTES SET OUT HIS PHILOSOPHICAL THEORY OF BRAIN FUNCTION, AN ENGLISH DOCTOR, THOMAS WILLIS, published the most detailed anatomical account of the brain to date. One of his discoveries still bears his name today.

Willis's seminal work was called *Cerebri Anatome*. It became the most important book in neurology for years to come, not least because it was the first place that the word "neurology" appears. Willis did not work alone; he acknowledged the work of the anatomist Richard Lower, who helped with the dissections. The illustrations showing the many intricate details the pair revealed were drafted by none other than Christopher Wren, one of England's greatest architects—responsible for London's St. Paul's Cathedral among many other famous 17th-century buildings.

Loop of blood vessels

Willis tracked the arteries that supply the brain, and found a curious combination of vessels at the base of the brain. The two vertebral arteries snaking up the spine converge under the brain, forming the basilar artery. A few arteries branch off this, one of which connects to the internal carotids, which have arrived via the neck. A communicating artery then loops around between the carotids, closing a circle, the Circle of Willis. This arrangement of vessels has built-in redundancy. If one vessel fails, the blood can always find another way: The brain is one tough organ.

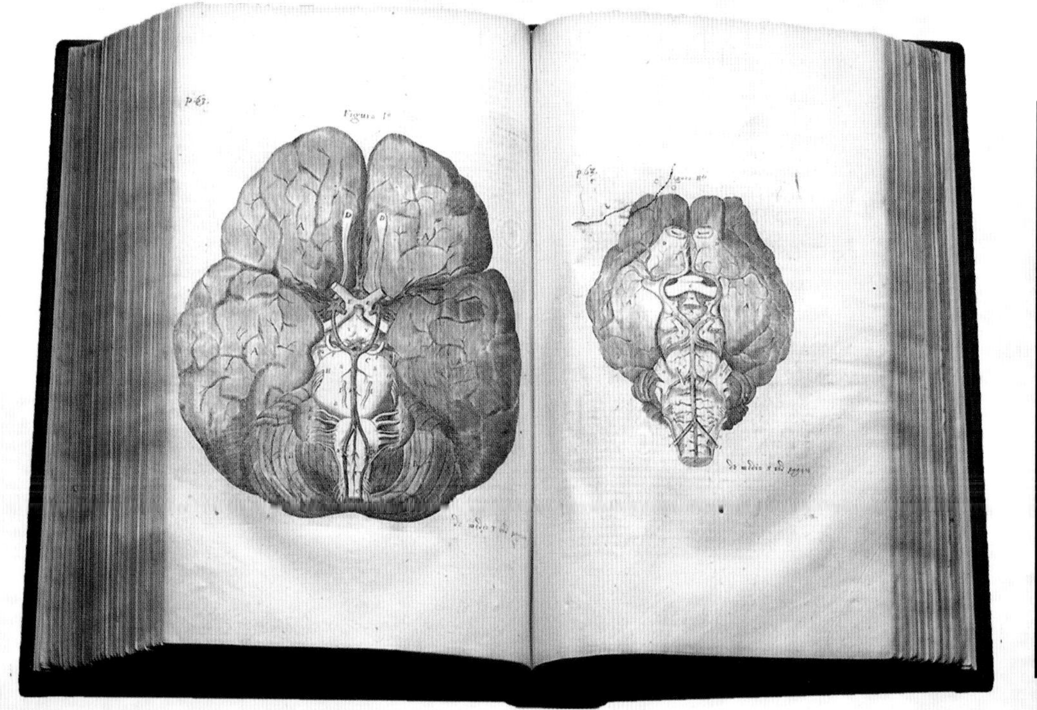

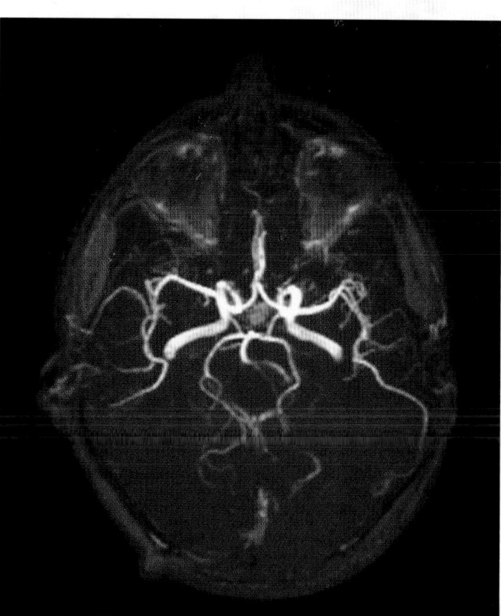

The Circle of Willis as shown by the man himself, and in a modern angiogram.

22 Functional Anatomy

THOMAS WILLIS WAS NOT ONLY INTERESTED IN MAPPING THE PHYSICAL STRUCTURE OF THE HUMAN BRAIN. He wanted to understand what each part did. His ideas moved neurology on from the days of ventricular spirits and ushered in the modern science.

Thomas Willis, the father of neurology, suggested the brain acted like the king of the body, sending out orders to exercise ultimate control.

Until Willis and friends published *Cerebri Anatome*, most people with an interest in the brain applied their common sense. Common sense was what Avicenna had said drove the brain: The senses delivered fragmentary details of the outside world to the lateral ventricles toward the front of the brain—close to the main senses. These ventricles then merged this sensory data into one single picture of reality: The common sense. The common sense then passed to the middle ventricle, where intellect and reason figured out what to do with it. Then it was transferred to the rear ventricle where it was confined to memory as a record of that moment in time.

GYRI AND SUCLI

The surface of the human brain is folded with plump, rounded ridges, known as gryi (singular gyrus), which weave around to form a relatively uniform pattern that is seen in every normal brain. The "valleys" that separate one gyrus from another are called sulci (singular, sulcus). Willis proposed that each gyrus was responsible for a certain higher function. As part of the theory, Willis suggested that the gyri received sensory information from a dense region under the cerebral hemispheres, which he called the striatum. It was the striatum that also controlled voluntary movements.

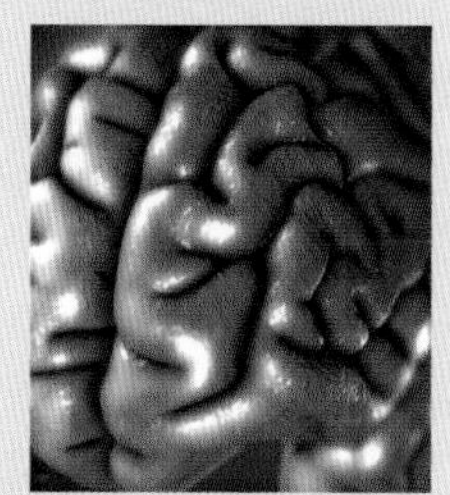

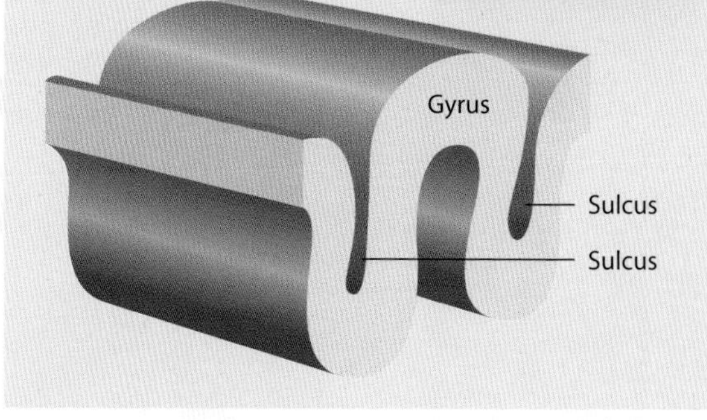

Using reason

Willis had other, entirely different ideas. He arrived at them by comparing the human brain with those of other animals, and from observations of brain injuries he saw while practicing as a doctor in Oxford, England.

Willis suggested that the higher regions of the brain were the location of the higher human faculties—memory and imagination. His reasoning was that this was the largest part of the human brain, far bigger than the equivalent structures in animals, and so must be the part that gave us our intelligence.

Lower down

The lower parts of the brain, clustered above the spinal cord, were involved in the more basic functions, although more vital in every sense of the word. Willis called this region of the brain the cerebellum—meaning the "little brain" and contrasting with the cerebrum, which refers to (most) of the rest. However, Willis's use of cerebellum was applied more loosely than it is today and probably comprised other hindbrain structures, such as the pons and medulla, which are considered distinct today. Willis differentiated the "gray matter" of the cerebrum, and the "white matter" which were fibers that spread from the brain to the body—carrying animal spirits with them.

23 St. Vitus's Dance

VITUS, AN EARLY CHRISTIAN WHO WAS MARTYRED INTO SAINTHOOD DURING THE ROMAN PERSECUTIONS, IS THE PATRON SAINT OF DANCERS. St. Vitus's Dance was an early name for another movement disorder, another "chorea." Today, we know it as Sydenham's chorea for the work of a 17th-century doctor.

The data from the 17th century shows that St. Vitus's Dance was more common in women and young people, a fact born out by modern studies.

In 1686, English doctor Thomas Sydenham wrote a description of the chorea that would later take his name: "This is a kind of convulsion, which attacks boys and girls from the tenth year to the time of puberty. It shows itself by limping or unsteadiness ... The hand cannot be steady for a moment ... He makes as many gestures as a mountebank (an old name for a con artist)." His suggestion was that a "humor had fallen on the nerves." He prescribed bleeding from the arm and plasters applied to the feet. These seemed to work. Everyone recovered.

Disease of the young

Later research revealed much the same thing. Young people below the age of 20 were the most likely sufferers of Sydenham's chorea. They had trouble walking, and writhed so much while lying down that they often threw themselves out of bed. However, once they fell asleep, the convulsions ended. However, the disorder was always temporary, generally lasting a few weeks, perhaps months. Sufferers were known to relapse, but few if any died.

By the turn of the 20th century, the cause of Sydenham's chorea had been shown to be a bacterial infection. Its particular effects were due to inflammation of the motor cortex, a region of the brain associated with voluntary movements. The motor cortex had only recently been discovered, and the link to this temporary disorder was further evidence of its function.

Sydenham's chorea produces a range of involuntary movements. Today, it can be treated with penicillin.

24 The Nature of Knowledge

HAND IN HAND WITH MEDICAL RESEARCH, THE NATURE OF THE BRAIN WAS ALSO QUESTIONED BY PHILOSOPHERS. Englishman John Locke had a real doozy. If knowledge could be moved from one brain to another, did the person travel with it?

John Locke had the good fortune to be a doctor as well as a philosopher. In 1689, his *An Essay Concerning Human Understanding* considered what would happen if two people woke up one day with each other's memories? Their bodies would be unchanged as would their circumstances but they would no longer remember how they had ended up in that particular bed in that house. The history of their body was unknown—but they remember the history of another body, which is somewhere else. So, in the end, who are they?

BLANK SLATE

John Locke was the founding figure of the Empiricist school, which proposes that knowledge comes from experience. Therefore, the human brain, according to Locke, was a tabula rasa, or blank slate, at birth. It was entirely empty but rapidly filled. Does modern neuroscience agree?

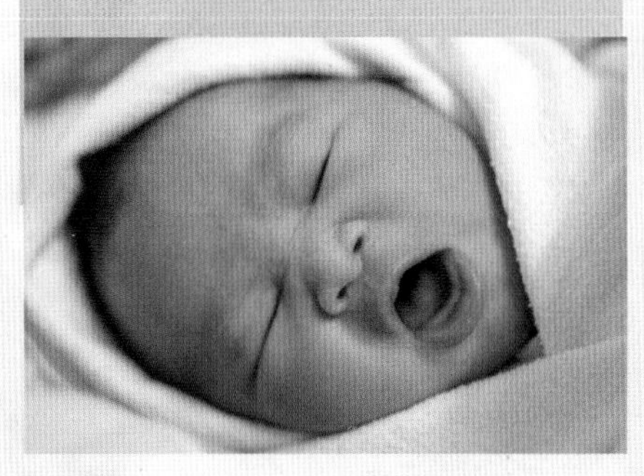

25 Idealism

GEORGE BERKELEY WAS ANOTHER PHILOSOPHER WHO HAD AN IMPACT ON THE PASSAGE OF BRAIN SCIENCE. While Locke proposed that the brain was filled by experience of the world, Berkeley suggested that the brain was the only place that the world existed!

Bishop Berkeley wondered how the mind made our perceptions.

Berkeley was an Anglo-Irish bishop. He published several works on the subject he called "immaterialism." His central philosophy is sometimes summed up as "to be is to be perceived." That is a little disingenuous, since Berkeley did not doubt that an external physical world existed, but his problem was whether we could ever find anything out about it. Berkeley said that all we have are mental impressions of the external world provided by the senses. There is no evidence of anything else. And that means that the world does not cause us to understand it in a particular—or truthful—way. Instead, we project our own experience onto it, and causes and effects we might observe are down to our actions, not that of nature itself. This viewpoint persists in how we understand human consciousness—does that exist only because we have perceived it?

26 The Optic Chiasm

ONE OF THE LARGEST AND MOST OBVIOUS NERVE STRUCTURES ASSOCIATED WITH THE BRAIN IS THE OPTIC CHIASM. It is formed by the two optic nerves meeting in a thick bundle underneath the brain. Known since antiquity, the bizarre nature of this junction was finally revealed in the mid-1700s.

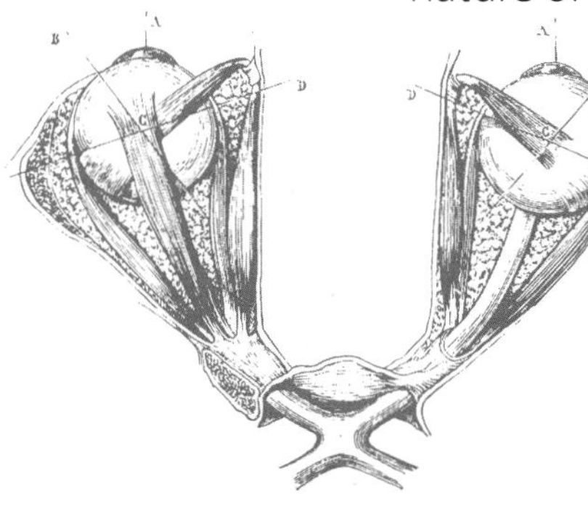

The optic chiasm is located under the hypothalamus.

Galen and other proponents of the extramission theory of vision saw the chiasm as the source of the eyes' "fire." However, as a more reasoned theory of vision took shape, enquiry changed to the route taken by sensory information through the junction. The word "chiasm" means "crossing" but until the 18th century it was assumed that the optic nerves did not actually cross. Isaac Newton, the author of *Opticks* in 1704, suggested that there was a partial crossing (to many people's confusion). In 1719, Giovanni Battista Morgagni provided evidence of it by recording how a brain-damaged person can lose vision in one-half of both his eyes. This is because signals from the field of vision nearest to the nose cross to the opposite side of the brain at the chiasm. Vision from the outer, temporal part of the eye does not switch sides. This arrangement helps make the flat images from each eye into a 3-D binocular image.

27 Animal Electricity

UNTIL THE 1780S, THE MECHANISM BY WHICH SIGNALS TRAVELED ALONG NERVES HAD MOVED ON LITTLE FROM THE ANIMAL SPIRITS of old. Then an Italian anatomist made an accidental discovery.

Luigi Galvani had followed his father into the medical profession. As part of his training in surgery he developed an interest in anatomy and, eventually, he became a full-time anatomist at the University of Bologna. After nine years of academic research, Galvani made the accidental discovery for which he is remembered. It would not only revolutionize anatomy but be a turning point for neuroscience and even physics.

Galvani had hung up pairs of recently severed frogs' legs on a wire fence to dry out. The fence was iron while the hook was copper. The fresh frogs' legs started to twitch—even spark according to some accounts! Galvani found he could recreate the twitching with

Galvani's diagram of frog's legs moving by "animal electricity."

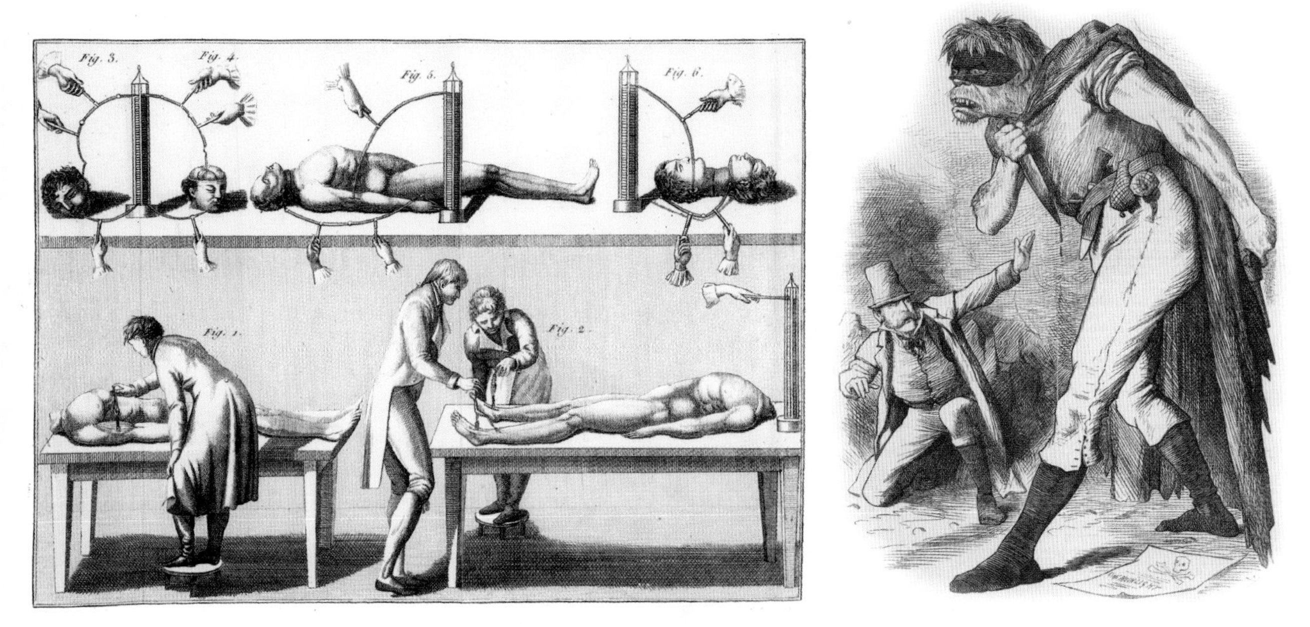

Giovanni Aldini's demonstrations of the power of electricity on the dead body has lived on in the popular imagination as Frankenstein's monster.

electrical charge, showing that living (or at least recently dead) muscle was stimulated by what he termed "animal electricity."

Nervous energy

The Italian investigated further to find that he could repeat the phenomenon by using a curved wire—again made from copper and iron—to connect the exposed nerve to the end of the leg. Unknown to Galvani, he had formed a primitive circuit that allowed his "animal electricity" to flow along the nerve and into the muscle, causing it to contract. Galvani reported his technique worked equally well on larger mammalian subjects and he even showed that the human body could be used as part of the circuit. Galvani believed that he had found some curious, animated quality of animal tissue. However, other scientists later showed that electricity did not require animal tissue, and in fact it worked much more effectively without. Nevertheless, electricity and the workings of the body became the focus of future researchers, who investigated its use as a therapy and its relationship to nervous and muscular function.

Reanimation

The link between electricity and nerves was shown to great effect by Galvani's nephew, Giovanni Aldini. In the early 1800s, he turned animal electricity into a traveling road show. He toured Europe zapping the nerves of recently executed convicts. Aldini was able to open and close the mouth on a severed head, make it blink, and swivel its eyes. Many future neuroscientists were inspired by stories of Aldini's macabre performances, as was Mary Shelley, the author of *Frankenstein*, a story of a monster reanimated by electricity.

FRANKENKITTY

In 1817, Karl August Weinhold had a go at a real-life Frankenstein's monster—only in his version he used a cat. The German scooped out the brain and spinal cord of a recently dead cat. He then poured a molten mixture of zinc and silver into the skull and spinal cavity. He was attempting to make the two metals work like an electric pile, or battery, inside the unfortunate cat, replacing the electrical signals of the nerves.

Weinhold reported that the cat was revived momentarily by the currents and stood up and stretched in a rather robotic fashion!

28 Phrenology

THE CONCEPT THAT EACH PART OF THE BRAIN HAD A DISTINCT FUNCTION REMAINED FIRMLY ESTABLISHED, but progress in showing which bits did what was slow. By the 19th century, a German suggested that looking on the outside of the head was all that was needed.

The man in question was Franz Joseph Gall, a doctor who is the founding figure of phrenology. That term means "the study of the mind," and it was an important contribution to the field of cortical localization—the attempt to map function onto locations in the brain. Phrenology's chief contribution was that it was total nonsense, and once exposed as such, helped to steer neuroscience toward more constructive areas of enquiry

Franz Joseph Gall is the founding figure of phrenology. His work had a large influence on 19th-century psychiatrists.

Shape of the head

The central proposition of phrenology is that the brain is a collection of self-contained "organs," each one devoted to a specific task. These organs developed during childhood and as they did so influenced the growth of the cranium, or skull. Gall began to lecture on his ideas in the Vienna medical school in the 1790s, but he fell foul of church authorities so he moved to Paris in 1802. It was here that phrenology caught the public imagination—where it would remain for many decades more.

Phrenology was easily grasped by the public. Many practitioners claimed to be able to predict the future prospects of a child by measuring his skull.

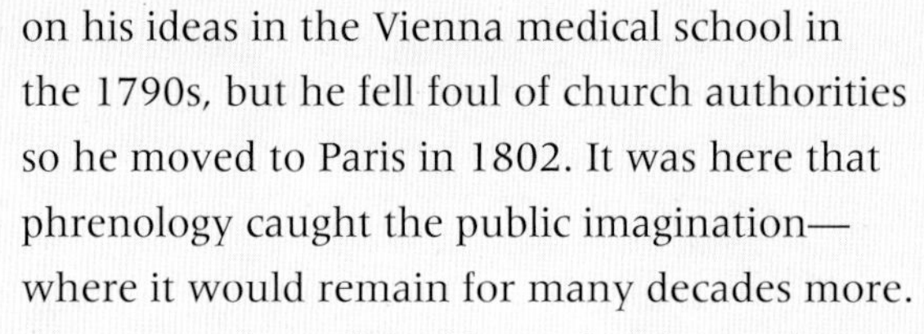

Amazingly, Gall had been developing his idea since the age of nine! He reports how he compared his academic abilities with one of his school friends. Gall was good at writing, but his friend was better at reciting poetry. The young Gall put this difference down to his chum having "cow's eyes"—in other words a slight bulge in the forehead. To Gall that was evidence enough that the verbal memory and speech centers were located at the front of the head. Someone else with a similar visage would also be good at talking and committing texts to memory. This was quite a clever observation for a child, but Gall did not grow out of it. To be fair, few people had published any other ideas about the localized brain functions. Gall's life work was set.

Animal and human

Gall began to build a map of the cranium. He did this using a number of techniques. Firstly, he studied the skull shapes of great thinkers, artists, and other high achievers. He looked for what made their crania different and this would indicate where their exceptional brain areas resided beneath. For the same reason, Gall also looked at the brains of criminals and the mentally ill. In total, he had the actual skulls of more than 300 people with some kind of extreme ability, plus he took castes of as many living subjects as possible. Of course, many of the functions of the human brain were shared with animals, and so Gall also studied the crania of a wide range of animals—each one known for a certain set of defining characteristics.

Gall mapped 27 regions on the human head—19 of which were shared with animals. He decided the region associated with destruction was located above the ear because this was pronounced in large carnivores. The urge to steal was found above this, because Gall had known pickpockets with large bulges here. Writers rubbed the side of their heads—and Gall had seen this area enlarged in the busts of many poets. This area was given over to ideas and ideals. Religious people had bulges on the very top of the head, so Gall put the ability to venerate there.

Gall said that the brain's hemispheres duplicated each other and worked in symmetry. He assumed that brain injuries to one side disrupted the balance and caused a loss function. However, despite its general public popularity, the science of phrenology was questioned from the outset. One of the strongest arguments against it was that functions could recover, even if its part of the brain was removed.

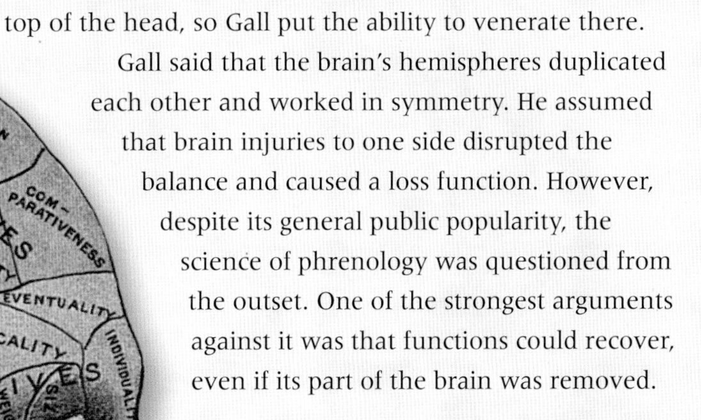

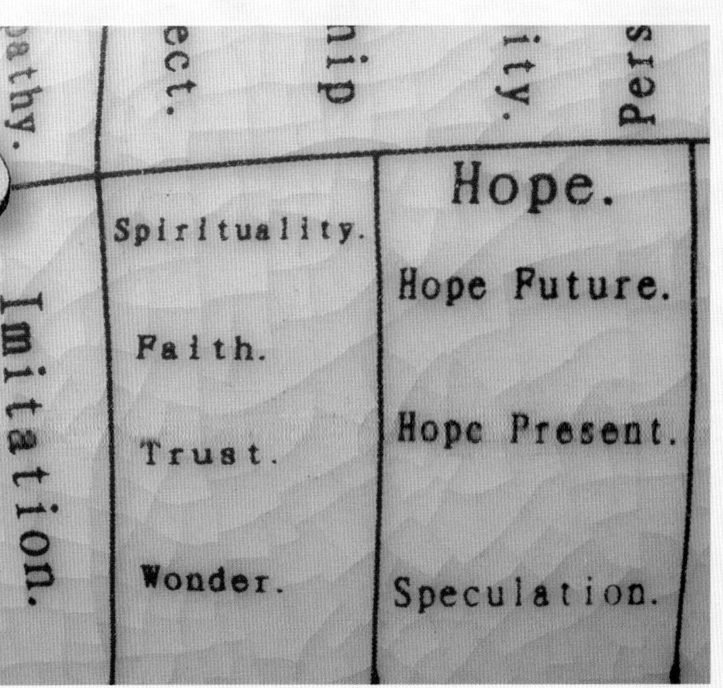

Gall's phrenology map placed higher qualities at the top of the skull, while lower, base appetites were controlled nearer the bottom. Gall left many empty spaces on the skull, but fellow enthusiasts filled it up over the following decades.

29 Parkinson's Disease

WITH THE WORLD POPULATION AGING, THE NUMBER OF SUFFERERS OF PARKINSON'S DISEASE IS SET TO INCREASE. This affliction, primarily of the old, is named for James Parkinson, who recorded it in 1817.

One percent of people over the age of 60 suffer from Parkinson's disease. By the age of 80, that figure is four percent.

Parkinson was a London surgeon with a writing habit. He wrote books on paleontology, politics, and chemistry. However, it is his *An Essay on the Shaking Palsy* that made his name, even though Parkinson's association with the disease would have to wait a century. In 1850, Germain Sée wrote a treatise on the different symptoms of Sydenham's chorea and the shaking palsy. He wanted to attribute some finding to Parkinson, but named him Patterson instead. Jean-Martin Charcot got it right when he coined the term "la maladie de Parkinson" in 1861, but the disorder was better known as *paralysis agitans*. Only when Leonard Rowntree, an American physician, went on vacation to England in 1912 did he reestablish the link with Parkinson, and reinvigorated investigations into the degenerative disease.

Shaking palsy

Parkinson's essay was based on observations of six patients, three of whom Parkinson was able to examine. The other three were seen on the street—and Parkinson was only able to stop and take a look at two, the third being seen from afar. Nevertheless, Parkinson managed to glean the defining features of the disease, which differentiated it from other movement disorders. Shaking palsy produces: "Involuntary tremulous motion, lessened muscular power ... propensity to bend the trunk forward ... The senses and intellects being uninjured."

Parkinson's patients were unable to report when the symptoms had started, and so he concluded that the disease was very slow in its progression. Parkinson also suggested that the disease was due to a disorder within the brain, rather than a problem with the nerves. That link was duly discovered in 1921 (*see* box).

BLACK SUBSTANCE

James Parkinson offered no cause for the shaking he described. Charcot suggested it was due to damp and melancholy, while American John Hutchling Jackson thought the tremors were due to the nerve impulses being intermittently blocked. It was Konstantin Tretiakoff, a Russian anatomist, who linked the disease with damage to the substantia nigra, a dark patch in the midbrain. Extensive autopsy work in 1921 showed than every Parkinson's sufferer had damage in this tiny region. Future research would reveal that the substantia nigra supplies the chemical dopamine to the basal ganglia, the region that controls movements. Parkinson's disease results from a disruption of this supply.

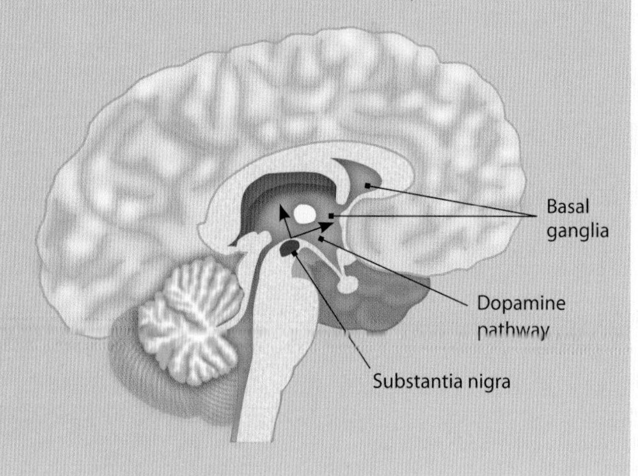

30 Bell-Magendie Law

THE FIRST STEPS IN REVEALING HOW DIFFERENT PARTS OF THE BRAIN AND NERVOUS SYSTEM ARE SPECIALIZED to perform certain functions were made in the early 1800s. The most important breakthrough was how the spinal cord worked according to a relatively simple rule.

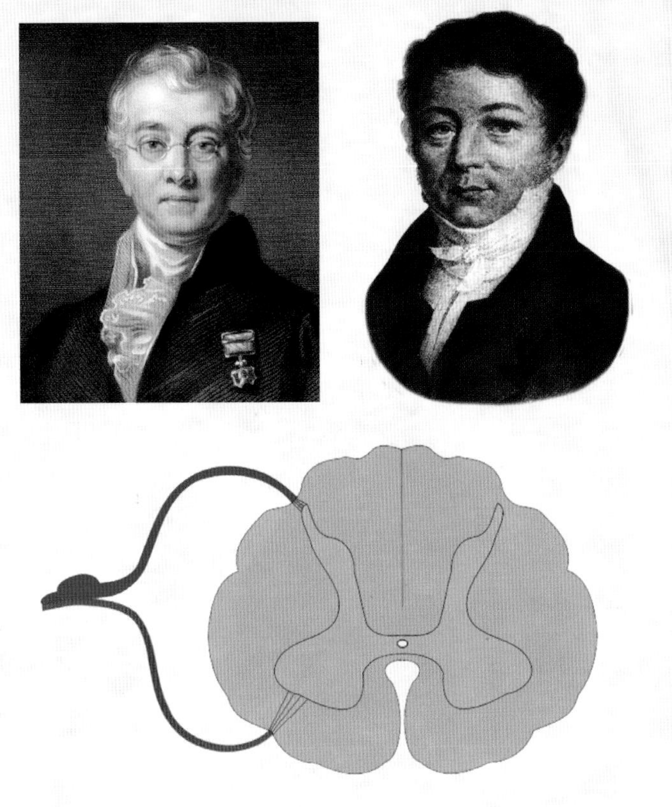

As Messrs. Bell (left) and Magendie showed, the sensory nerve (blue) brings signals into the back of the spinal cord, and the motor nerve (red) carries a response out the front.

In 1806, Jean César Legallois made the first proper discovery in cerebral localization—the drive to understand which bit of the brain did what. While experimenting with rabbits, the Frenchman found that if he severed the spinal cord below the eighth cranial nerve, the rabbit continued to breathe—and survived for several minutes at least. He had found the respiratory center, the part of the brain that controls breathing. It was located in the medulla, which until that point was regarded as the top end of the spinal cord. From then on, the medulla was seen as the bottom end of the brain!

Input and output

At the same time, two other researchers were studying the spinal cord as well. They were Charles Bell, a Scot working in London, and François Magendie based near Paris. Although working independently, both took puppies as their subjects, and they were both interested in how the spinal cord received information from the senses and how it sent out signals to the muscles. The general assumption was that all the nerves connected to the spinal cord—known as its "roots" due to their obvious resemblance to plants—were able to handle both kinds of information.

Bell's work was centered on dissection of dead specimens. In 1811, he found that the roots that emerged from the inner part of the spinal cord connected to muscles. He did perform vivisections (experiments on live animals) as well, but only on unconscious subjects to avoid causing the dogs pain. Therefore, he had little to say about the sensory nerves, which were largely inactive during his experiments.

In the 1820s, Magendie performed vivisections on puppies that remained awake. He confirmed Bell's discovery and also revealed that sensory nerves met the back of the spinal cord. In medical parlance, the motor nerves (the muscle-linked ones) are ventral, meaning they are on the belly side. The sensory nerves are dorsal—they are near the back. Despite considerable disagreement about who made the discovery, today this universal arrangement of the spinal roots is known as the Bell-Magendie Law.

SPINAL CORD

The spinal cord is the link between the brain and most of the rest of the body. There are 31 spinal nerves that obey the Bell-Magendie Law, each one with a motor and sensory root. Damage to the spinal cord leads to loss of movement and feeling below that point.

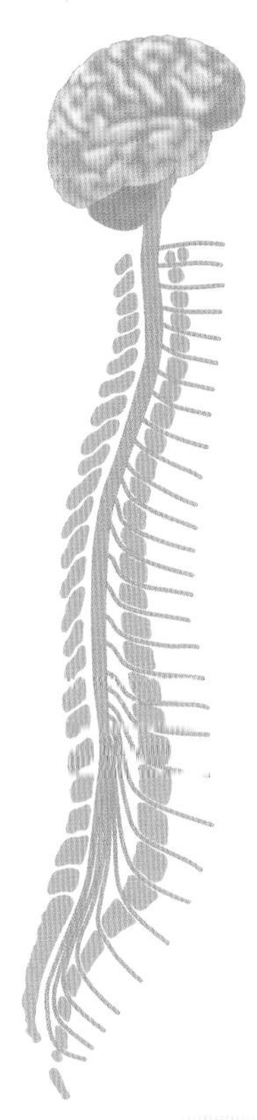

31 The Neuron

THE CONCEPT OF BODY CELLS DATES BACK TO 1665 WHEN ROBERT HOOKE FIRST SAW THEM THROUGH HIS PRIMITIVE MICROSCOPE. However, the role of cells in the nervous system had to remain a mystery until the 1830s, when microscope technology became powerful enough to look closely at the brain.

Jan Evangelista Purkinje contemplates his life's work.

Dendrites

Cell body

Axon

Robert Hooke had seen small distinct chambers in a sliver of cork wood. He likened them to the rooms used by monks for private study, and they have been known as cells every since. However, cork cells are huge compared to most of the cells in the human body, including those in the brain.

Pioneering neuroscientists could only see the gross anatomy of the brain and had to rely on lesions to glean information about how it worked. In other words, they made cuts into the living brain of animals and monitored the effects; they dissected the diseased brains of the dead to see what was making it unhealthy; or they waited for an unfortunate human to suffer an nonfatal brain injury and tested their abilities. In the 1820s, a new design of lens made microscopes powerful enough to look at the anatomy of the brain on the cellular level. What neuroscientists would find down at that scale provided a key to how the brain worked.

Branching cells

The groundbreaking microscope technology was the achromatic lens. This was able to focus light of any color, which created the clearest views of brain tissue yet seen, and at much higher magnifications. The brain was sliced with a knife or teased into fine layers using tweezers. As the microscope was brought into focus, researchers could see the cells of the brain—at least the largest ones—for the first time. The brain is so densely packed with cells it was impossible to tell where one cell ended and another began. However, three anatomists, working independently, were able to make out the first views of the brain cells.

THE NERVE CELL

Every neuron follows a general structure. The nucleus is in the cell body, which is surrounded by several branches. The shorter, more plentiful branches are the dendrites. A single large branch extends from the cell body. This is the axon.

The scientists were Christian Gottfried Ehrenberg, Gabriel Valentin, and Jan Evangelista Purkinje (pronounced "porkinyer"). It is Purkinje's drawings that gave the best report of the discoveries, as he studied the large cells found in the cerebellum. His diagrams look a little like tadpoles or fish, with twisting, branching tails. Further improvements in microscopes, coupled with novel dying techniques, would reveal that the cells had more than one tail, but several branches around a central cell body, or soma. Most extensions were very short and were described as protoplasmic prolongations. However, one of them was thicker and much longer. This was initially known as the axial cylinder. Today, the shorter extensions are known as dendrites. They are measured in millionths of an inch and are the receivers of the cell, collecting signals from its neighbors. The large one is the axon. This sends out signals from the cell and can be much longer than the dendrites and can be measured in feet!

"Deceptions of the senses are the truths of perception."

JAN PURKINJE

Cell theory

Two years after Purkinje's drawings became public in 1837, Theodor Schwann and others proposed the general cell theory of biology. This stated that every body had to have at least one cell in it, and that every cell arose from the splitting of another. However, no one could figure out if nerve cells, so bizarre compared to the rest, were included in this theory. In fact, were they even cells at all?

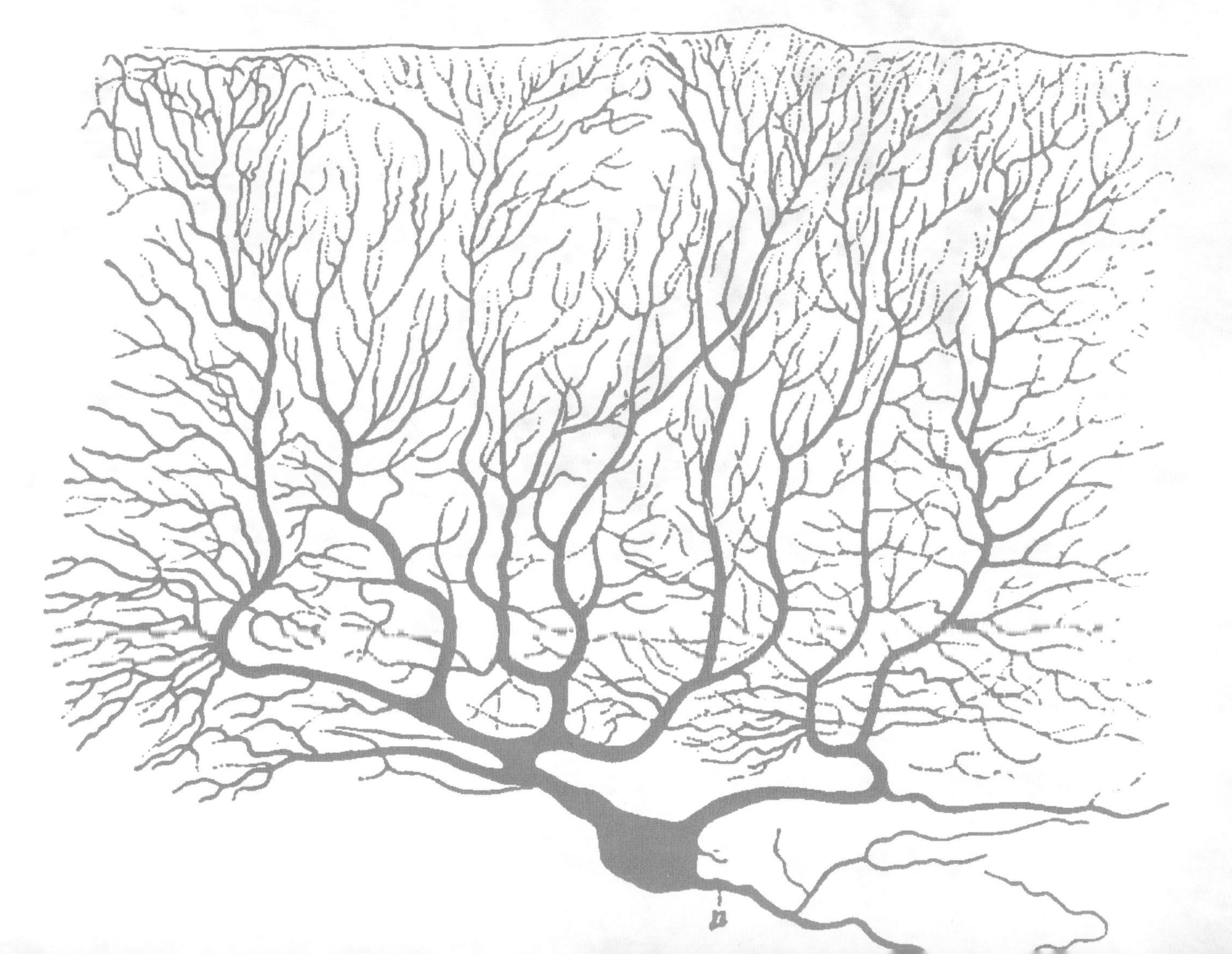

A hand-drawn diagram of a Purkinje cell, named for its discoverer. This drawing was made by Santiago Ramón y Cajal, who helped figure out the full structure of the neuron in the 1880s.

32 Anesthetics

DOCTORS HAVE LONG SOUGHT A MEANS TO REDUCE THE PAIN OF THEIR PATIENTS. VARIOUS CONCOCTIONS HAVE BEEN USED SINCE ANTIQUITY TO DULL THE SENSES, however, by the 1840s, a new raft of chemicals were discovered that could bring on unconsciousness. As well as being a great benefit to surgery patients, these anesthetics also shone a light on brain function.

The idea of anesthetic is as old as recorded history. The Babylonians relied on the juice of poppy seeds and stinking nightshade to reduce the perception of pain. Hippocrates and Galen were well aware of the sedative powers of opium. However, despite being effective painkillers, these medicines were relatively slow acting and were unpredictable. From these ancient times to the 19th century, any kind of surgical procedure was a horrific affair. Even if a patient could be sedated prior to beginning surgery, they were likely to awaken after anything more than the most innocuous incision. In the Middle Ages there were reported attempts to administer intoxicants with a sponge held over the mouth and nose of the patients as they appeared to stir. However, this technique fell from favor, probably because it caused many deaths.

Inhaled anesthetics can be used to keep patients unconscious and unaware of events during surgery.

CURARE

Found in American plant poisons used to coat darts and arrows, curare is a paralytic—it makes you paralyzed by relaxing your muscles. In the early 20th century, it was investigated as an anesthetic. However, the patients were able to feel every aspect of the surgery, but were incapable of doing anything about it. So, curare can be used to paralyze patients during surgery but anesthetics are needed to reduce their awareness.

Having a laugh

At the end of the 18th century, pneumatic science was the cutting edge of research. New "airs" or gases were being discovered, not least the likes of hydrogen, oxygen, and nitrogen. The Pneumatic Institution in Bristol, England, under Thomas Beddoes, was devoted to finding inhalable medicines among these airs. In the 1790s, Beddoes had encouraged tuberculosis patients to inhale the stench from a cowshed set up in their hospital, in the hope that it would protect the lungs! By 1799, he was working on nitrous air, and employed the help of a young Cornish scientist called Humphry Davy. Davy found that breathing in nitrous oxide made his body grow numb. It also lifted his mood and made him laugh a lot. The gas soon became known as "laughing gas." Davy and Beddoes held laughing gas parties for their friends, such as poet Samuel Taylor Coleridge, who was frequently self-anesthetized by opium.

Ether and chloroform were not without dangers. Various inhalers, like this one below, were invented to reduce the chance of overdosing.

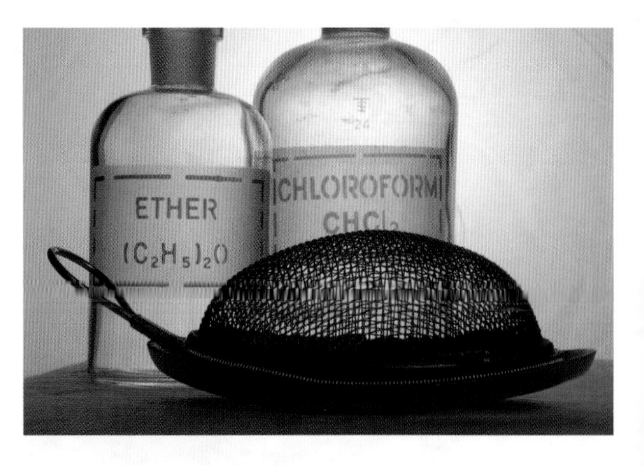

Davy's parties brought him into contact with the wealthy high society of London, where he began a career as one of Britain's preeminent scientists. He suggested that laughing gas might have medical uses. His idea was ignored at the time but in 1844 an American dentist volunteered to have his own tooth extracted while under the influence, and so began its use as a mild anesthetic.

However, Davy's assistant, Michael Faraday, had already discovered a more powerful anesthetic gas called "ether." As with laughing gas before it, ether parties or "frolics" became popular, and reports emerged that people became temporarily unconscious under its effects. In 1842, American surgeon Crawford Long, no stranger to the odd frolic himself, began using ether as a general anesthetic during minor operations. Later that year Scottish surgeon Robert Liston amputated the leg of an anesthetized patient in less than three minutes. The patient did not feel a thing!

In 1847, James Young Simpson began to experiment with chloroform, which had similar effects to ether (although eventually proved to be rather dangerous). The question was: Did these chemicals put people to sleep or do something else?

Interrupting consciousness

Anesthetics are different to sedatives, which make you drowsy, less responsive, and eventually fall asleep. Instead, they are also not strictly speaking painkillers either. They reduce awareness of pain and other stimuli. How they do that is still open to question. The drugs suppress the activity patterns of nerve cells in the brain, making them unable to form sensations and memories. In essence and in effect, they turn off consciousness.

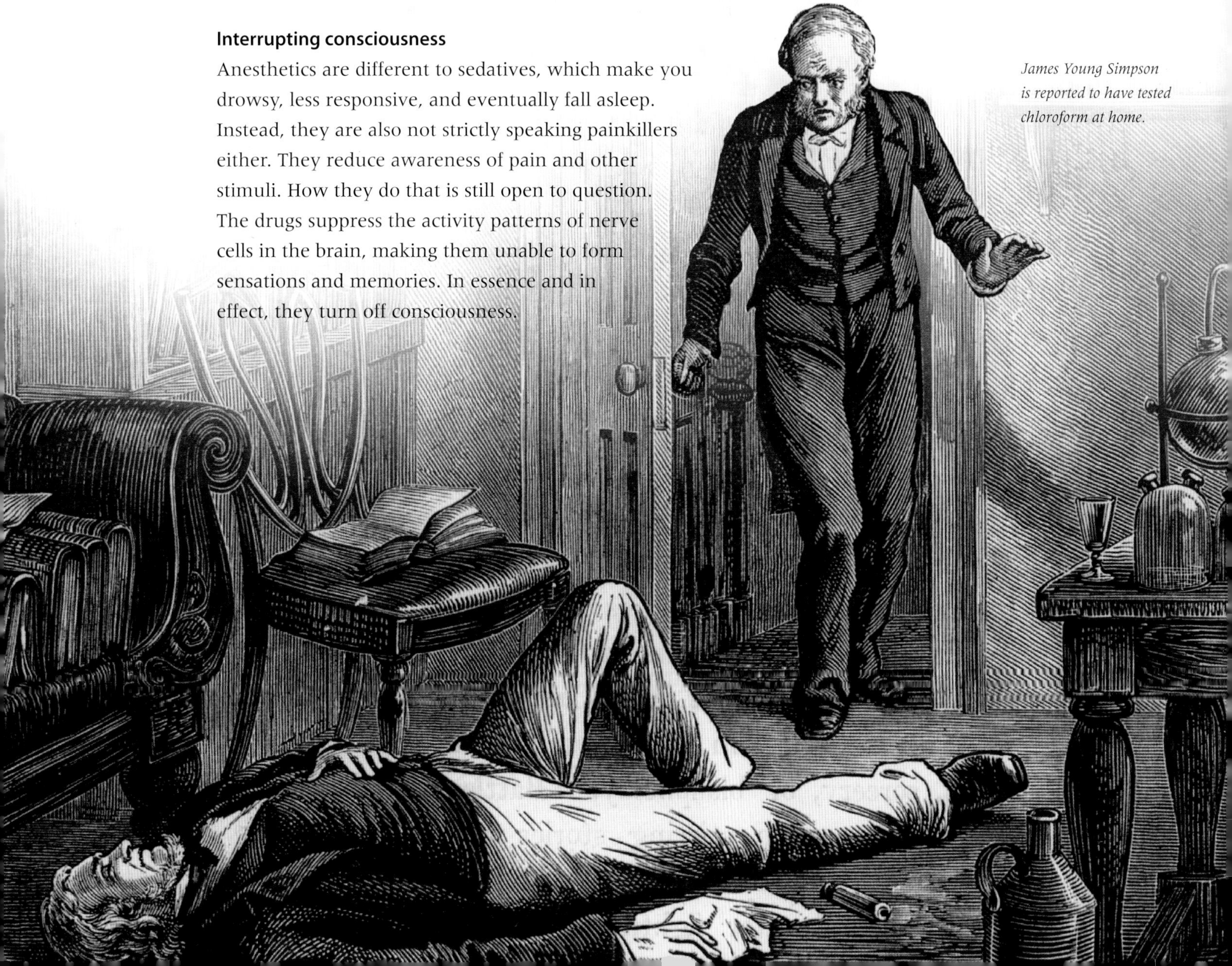

James Young Simpson is reported to have tested chloroform at home.

33 Phineas Gage

IN 1848, AN ACCIDENT TOOK PLACE AT A RAILROAD CUTTING IN VERMONT. An explosion fired a four-foot metal spike through the head of a workman. He survived the brain injuries but everyone was curious to know how they had changed him.

The unfortunate victim was 25-year-old Phineas Gage, foreman on the Rutland and Burlington Railroad. Gage was tamping down gunpowder in a hole bored into rock, using a long iron rod. A spark inside the hole ignited the powder, and in an instant the rod had been launched through Gage's left cheek and out the top of his head. He did not die, and when the medical help arrived, Gage said: "Doctor, here is business enough for you."

Phineas Gage shows off the tamping iron that changed his life. The only physical injury he suffered was the loss of his left eye. However, reports conflict on how his personality was changed.

Recovery

Gage drifted in and out of a feverish consciousness for two weeks, but recovered his strength remarkably quickly and was able to walk again within the month. Gage become a celebrity, and toured the country for a little while before settling down to work as a liveryman and stage coach driver. In 1860, he began to suffer seizures that made it impossible for him to work. He died a few months later.

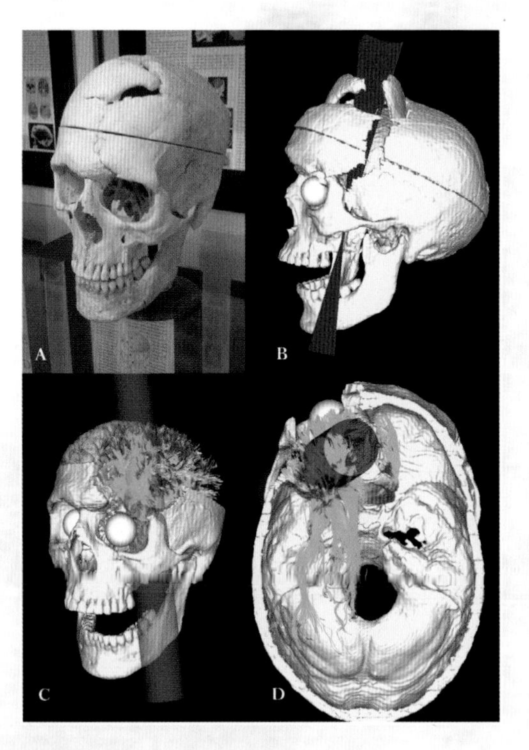

Phineas Gage's skull is on display at Harvard's medical school, Modern imaging shows just how severe his injury was.

No autopsy was made, but Gage's body was dug up in 1866. The skull shows that Gage's front lobe was severely damaged, and in accordance with phrenology, the dominant theory of the time, that would mean Gage would have lost his higher, human abilities, and would lack control over his more animal instincts. Popular accounts of Gage's final years often suggest he was a changed man. An anonymous poem summed it up: "A moral man, Phineas Gage, Tamping powder down holes for his wage, Blew the last of his probes, Through his two frontal lobes; Now he drinks, swears, and flies in a rage." It is now thought that Gage's extreme personality change has been exaggerated to tally with the presiding theories of neuroscience. If anything, Gage's accident disproved phrenological ideas, and showed that the injured brain did not always lose distinct functions, and had a remarkable ability to recover.

34 The Neurology of the Ear

THE ANATOMY OF THE EAR IS LIKE A TIMELINE, WITH ITS MANY PARTS NAMED FOR THE GREAT ANATOMISTS WHO FIRST DESCRIBED THEM OVER THE CENTURIES. By the 1850s, the structure of the ear had been investigated right down to the cellular level using a dye derived from boiled insects.

The gross anatomy of the ear, inside and out, was first described by Celsus in the first century C.E. The auditory ossicles, the tiny bones that sit in the inner ear, were clearly described by Vesalius in the 16th century. Many of the names we use for the ear's structures come from Gabriel Fallopius, a student of Vesalius, who is better remembered for his work on the ovary and uterus. It was Fallopius who coined the term cochlea, for the shell-like whorl deep inside the inner ear. A contemporary, Bartholomeo Eustachio, is remembered through the name Eustachian tube, which connects the inner ear to the throat.

All these advances led to several theories of hearing, all of which traced the sound wave to the cochlea but then resorted to guesswork. In 1851, Alfonso Corti developed carmine cell stain, using a red dye produced from South America's cochineal bug. This stain showed up distinct cells within the cochlea, and Corti saw hairlike structures lining the inside. These "hair cells" were the nerve endings that converted sound to brain signals. However, it would take another 80 years to figure out how they did it.

TRANSDUCER

The ear is a natural transducer, a device for changing wave energy into different forms. Sound is a pressure wave in the air, and its pitch relates to its wavelength. The ear converts this pressure wave into mechanical motion, then waves in liquid, and finally an electrical signal that travels to the brain.

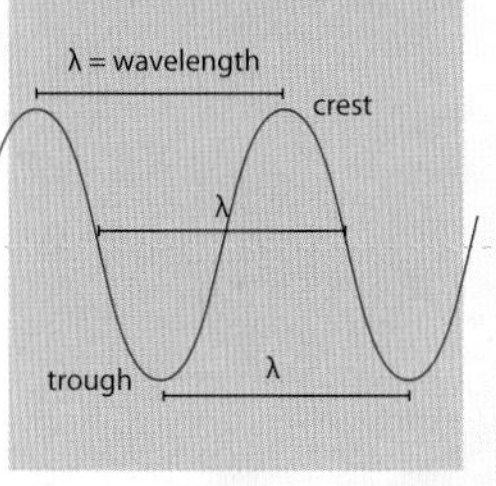

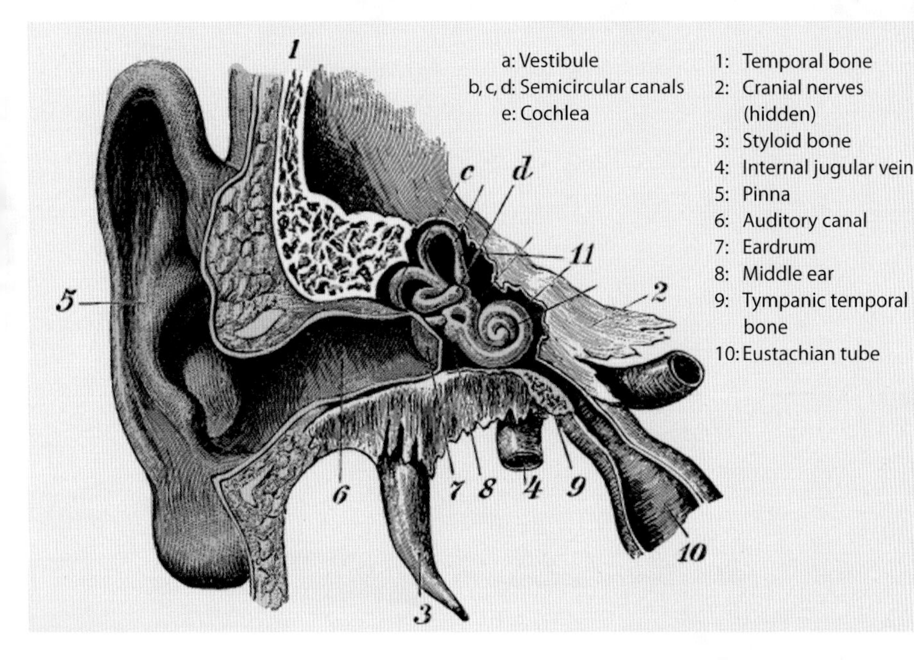

a: Vestibule
b, c, d: Semicircular canals
e: Cochlea

1: Temporal bone
2: Cranial nerves (hidden)
3: Styloid bone
4: Internal jugular vein
5: Pinna
6: Auditory canal
7: Eardrum
8: Middle ear
9: Tympanic temporal bone
10: Eustachian tube

EAR HERE

The ear is divided into the three parts: the outer, middle and inner ears. The outer ear begins with the external pinna, a dish-like flap that collects sound waves and directs them into the auditory canal. The wave continues to the ear drum, a flap of skin that converts the air wave into a mechanical vibration. From here the sound information has moved into the middle ear. Three tiny bones (not seen on this diagram), transmit this rhythm to a complex fluid-filled chamber, which makes up the inner ear. Ripples extend through this chamber. The most significant part is the shell-like cochlea, which is lined with minute filaments. These waft in sync with the ripples, creating an electrical nerve signal that is sent to the brain.

A hand-drawn ear diagram from the 19th century reveals most of the features of the ear.

35 Olfaction

THE BIOLOGICAL TERM FOR SMELLING IS OLFACTION. NOT ALL ANIMALS DO IT WITH A NOSE, BUT THE RELATIONSHIP BETWEEN THE NOSE AND THE HUMAN BRAIN has been under investigation since the days of Classical Greece. By 1856, researchers had located the connection between the two.

HENNING'S PRISM

Many attempts to classify smells have been attempted. The most successful is Henning's prism, formulated by Hans Henning in 1916. Every odor has a place within the prism according to its six "dimensions." Although good for helping to describe smells, the prism is limited. The latest research suggests that the human nose could differentiate a trillion distinct smell mixtures.

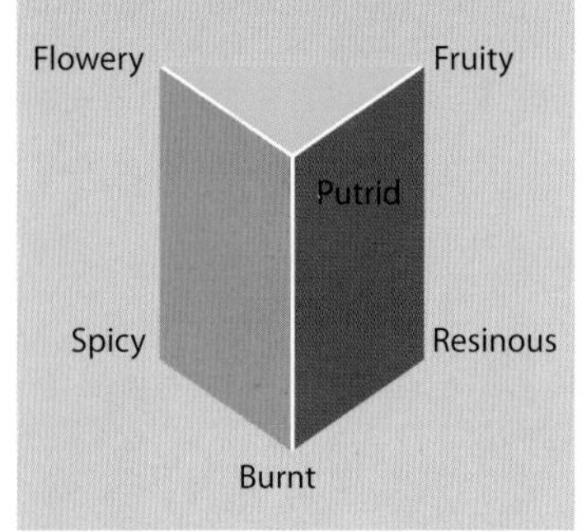

In ancient Greece it was thought that the sensations of smells were produced by smokes and airs sucked into the brain through the nose. Galen's explanation was that odors seeped through porous bones and were transported to the ventricles.

Andreas Vesalius put a stop to all that and described in great detail the olfactory tract and olfactory bulb. The latter is a mass of nervous tissue on the underside of the brain. It is against the upper surface of the ethmoid bone. The olfactory tracts is a bungle of axons that leads back from the bulb into the main body of the brain.

The search for receptors

Severing the olfactory tract was shown to stop the sense of smell. Moritz Schiff did that to four of five newborn (and still blind) puppies and showed that the ones without the nerve could not find their mother's teat! However, the location of the nasal receptors was still something of a mystery. Corti's staining technique was put to use once more to reveal the fine cellular structures of the nasal cavity. Starting in 1856, researchers such as Max Schultz and Conrad Eckhard, began to describe a network of axons that ran from different types of receptor cells in the nose through the ethmoid bone to the olfactory bulb. Each receptor cell has a dozen hairlike cilia, which are involved in collecting chemicals from the breath. It was quickly accepted that these nerve endings supplied signals to the olfactory bulb and then onward along the tract to the olfactory cortex, the smell center of the brain. Later research showed that this is in the temporal lobe, and unlike all other senses, information from the nose (and tongue) does not pass through the thalamus, the brain's information hub.

ON THE NOSE

The nose is a cutaneous sensor, using a patch of skin to pick up chemicals floating in the air. The skin in this case is the lining of the nasal cavity (p), where it is kept damp and clean, and connected to the air by the nostrils. Air rushes in through the nostrils and passes over the thumbnail sized patch of olfactory receptors on the roof of the cavity. There are more than 1,000 types of receptor, each capable of detecting several types of odor.

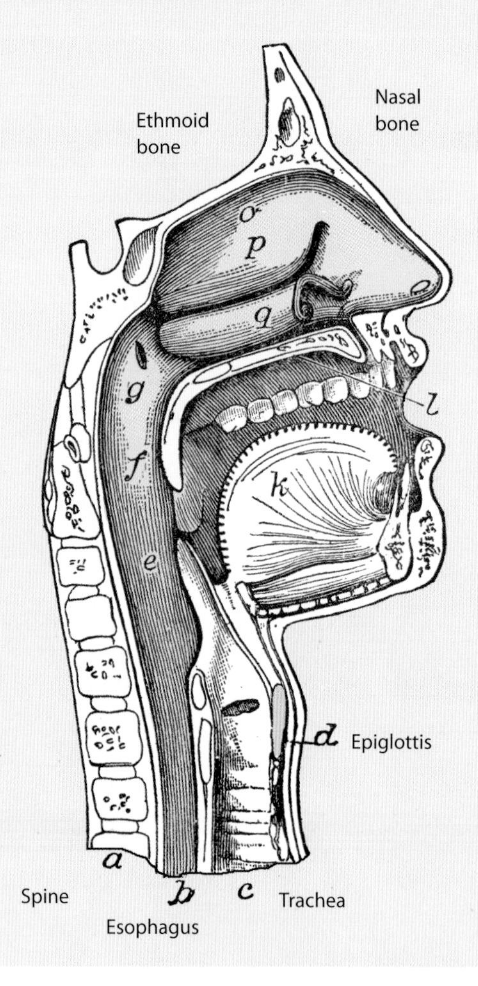

36 Glial Cells

German Rudolf Virchow's initial description of glial cells was "nervenkitt" meaning "nerve cement"—an idea about their function that persisted for 150 years.

THE DISCOVERY OF NEURONS WOULD REVOLUTIONIZE THE UNDERSTANDING OF THE BRAIN. In 1858, a new kind of brain cell was discovered. Known as glia, or glial cells, they were regarded as helpers for the neurons, a role that has been recently called into question.

In the 1830s, Theodor Schwann, the founding figure of cell biology, was driven to consider the universal characteristics of cells through his work on newly discovered neurons. Whatever animal he looked at, the nerve cells looked the same. One of their features was small cells that surrounded the axons. These are known as Schwann cells today. They are involved in coating the axon in a fatty sheath, known as myelin, which helps to transmit the nerve signal. Schwann cells are now seen as one of the many glial cells. Glial means "glue," the idea being that the cells held the neurons together.

Star cells

Glial cells were first differentiated by Rudolf Virchow in his 1858 book, *Cellular Pathology*. Many glial cells are star-shaped, and the most branched are known as astrocyctes because of this. The most common type are the oligodendrocytes, which have fewer extensions. They have connections with the axons and other parts of the neurons and maintain the chemical balance around the cell and boost blood flow when the neuron is active. Smaller microglia function as the brain's immune system. Glial cells do not produce electricity, but communicate chemically. Recent research is looking into how a chemical glial network may work alongside the electrical web of neurons.

NEURONS AND GLIA

There are many more types of neuron (a few are shown here) than glial cells. By most estimates, there are equal numbers of both cell types in the body, but glial cells outnumber the neurons three to one in the cerebral cortex.

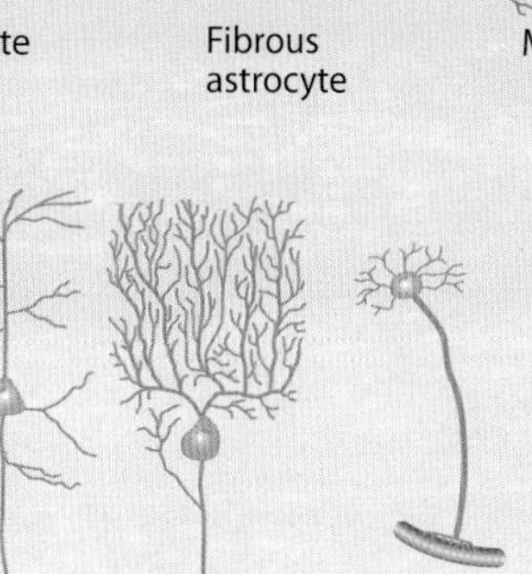

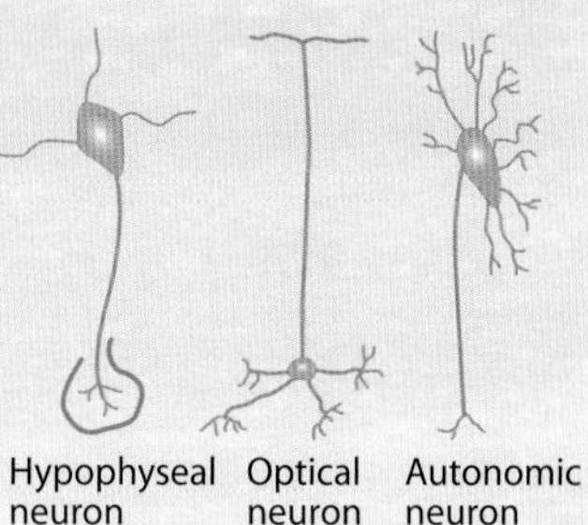

37 The Speech Center

FEW PEOPLE HAVE PART OF THE BRAIN NAMED AFTER THEM, BUT THAT HONOR HAS BEEN BESTOWED ON PAUL BROCA. In 1861, the Frenchman discovered the strongest evidence yet that body functions were controlled in distinct areas of the brain.

Paul Broca combined medicine with a career in politics and an interest in anthropology.

Broca's discovery was all the more thrilling because it concerned not the primitive functions of breathing or moving, but that most human of ability: speech.

Franz Gall and the other phrenologists had proposed that fluent speech was governed by organs at the front of the brain, above the eyes. Supporters of this theory showed off cases where soldiers wounded in that region sometimes lost the ability to speak—although they frequently ignored the wounded men who could talk. But a movement was growing in neuroscience to do away with the practice of measuring the shape of the skull to ascertain brain function. Instead, the consensus was to perform autopsies on the brains of people, recently dead, who had shown neurological symptoms. The hope was abnormalities in the brains would reveal which parts of the brain did what.

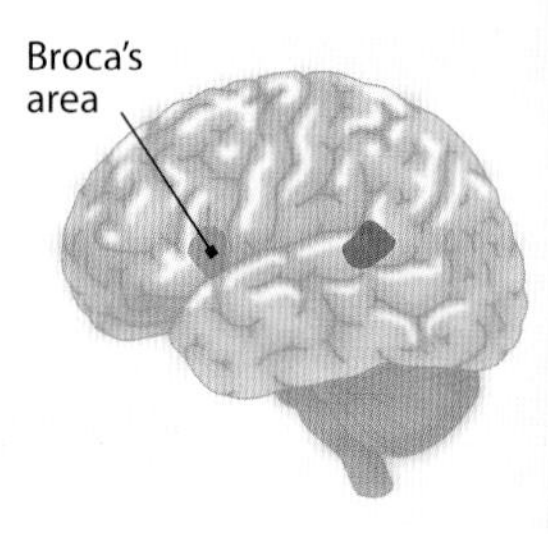

APHASIA

Speech loss, or aphasia, is one of the most obvious signs of brain injury. It is most common after a stroke or a head injury, and is mostly associated with damage to the left side of the brain, although not only to Broca's area. Damage there results in nonfluent aphasia where the speech is halting. Other sufferers with aphasia may be able to speak, but are unable to think of what to say. Speech therapy can sometimes solve the problem.

Monsieur Leborgne

In 1861, Paul Broca was working at the Paris medical school when a severally disabled patient was bought to his surgical ward. The man, known as Monsieur Leborgne, had been in various hospitals for 21 years. He was now paralyzed down his right side, suffered epileptic seizures, and could only say one word: "tan." He died six days after being first examined by Broca; there was nothing anyone could have done. An autopsy on Leborgne's brain showed a severe damage to the left frontal lobe.

A few months later a stroke patient came to the ward. His name was Lelong, and he was able to say five words: "Yes," "no," "three," "always," and "lelo"—as near to his own name as he could manage. At autopsy, his brain was damaged in the same place as Leborgne's. This region is now called Broca's area. Broca had been able to show that even though the tongue and other mouth muscles were able to move perfectly well, proper, fluent speech was impossible if there was damage to this one part of the brain.

38 Taste Buds

DESPITE ITS GREATER SENSITIVITY, THE HUMAN SENSE OF SMELL IS LESS SIGNIFICANT THAN ITS BEDFELLOW, TASTE. WE MISS A DECLINE IN TASTE a lot more than a loss of smell. In 1867, microscope research revealed the taste buds on the tongue, although at first they were known as taste bulbs, taste cups, and even taste hairlets.

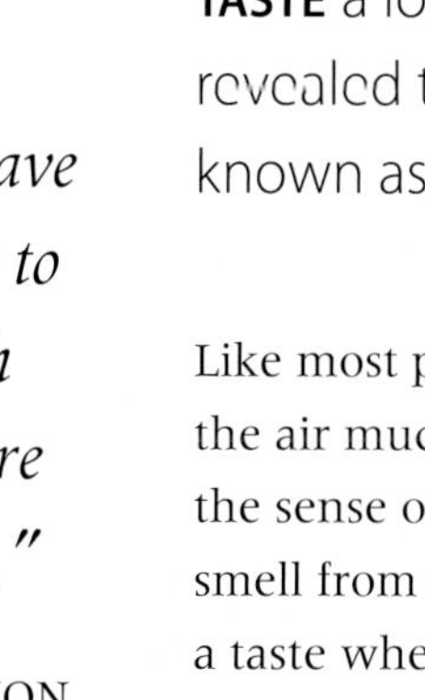

A rather gruesome drawing of a tongue shows its different textures. Large circumvallate papillae dominate the rear of the tongue, while foliate ones are most common along the side. Filiform and, to a lesser extent, fungiform papillae dominate most of the upper surface.

"Brute animals have the faculty to distinguish flavors more accurately."

ALBRECHT VON HALLER

Like most primates, humans do not live in a word of scent. They do not sniff the air much. Instead, the sense of smell's most significant role is to boost the sense of taste. In fact, much of what we "taste" is actually from what we smell from fumes in the throat as we chew food. (Sight is also involved—try to define a taste when blindfolded and hold your nose. Everything starts to taste the same!)

Taste sensation

The ancient view was that taste particles soaked into the tongue and were transported to the ventricles; Avicenna said that only materials that mixed with saliva could make this journey. By the 18th century, it was thought that the tongue could detect four chemical flavors: sweet, salty, bitter, and sour (two more have been added since). Close examination of the tongue showed papillae, proud bumps on the surface. There are four types: Fungiform (mushroom shaped), filiform ("hairy"), foliate (leaf shaped), and large domed circumvallate papillae. In 1867, the taste receptor cells were located within the papillae. They were found in bulb-shaped "buds" numbering from a few dozen to 2,500, depending on the type of papilla. The average in humans is 250 per papilla; circumvallate have the most, foliate the least, while filiform have none at all. Taste receptors were seen to resemble nasal ones, and three cranial nerves carried signals to the gustatory cortex located just below the olfactory one in the temporal lobe.

Gustatory hairs
Tongue tissue
Support cell (purple)
Gustatory receptor cell (blue)
Connection to cranial nerve

TASTE RECEPTOR

Taste buds are found on the tongue, the soft palate, cheeks, and throat. Their gustatory receptor cells are embedded into the surface and tipped with hairlike projections. These hairs bond to chemicals in the food, and that creates a nerve impulse. It is often said that taste buds at different parts of the tongue are sensitive to certain tastes. This is not true—all parts of the tongue can pick up all kinds of taste.

39 Neuroscience and Racism

THE ROOTS OF RACISM RUN DEEP BUT THEY WERE FERTILIZED BY EARLY ATTEMPTS to classify humanity into races, and one of the key characteristics measured was the size of the brain.

It comes as no surprise that when European scientists embarked on the new field of enquiry called anthropology—the study of humans—their findings were skewed by the ignorance and prejudices of the time. The same can be said of all scientific hypotheses. However, what is startling is how these unsubstantiated and frequently refuted results have persisted in the modern imagination, and are still cited to promote racist views.

Nott and Gliddon's drawings from 1868 compare the idealized features of Renaissance sculpture with the anatomy of apes. Their overtly racist study gave a pseudoscientific veneer to the claims of white superiority, which have persisted long in the public imagination.

Science meets racism with the work of Carl Linnaeus, a man famed for his system for classifying organisms. Linnaeus is reported by some to be a rather unpleasant person, but his views were by no means unusual in the 18th century. Linnaeus classified the human species as *Homo sapiens*, but also divided it into four races: *europaeus*, *americans*, *asiaticus*, and *afer*. Linnaean classifications were based on the outward appearance of organisms (biologists do it a little differently now), and Linnaeus used this logic on humans: The *europaeus* race wore clothes, while the members of *afer* did not! Europeans were educated, just, and disciplined; Africans did not share these traits. In 18th-century Europe, everyone would have agreed.

Skewed and caricatured skull anatomy was a frequent feature in the debate over racial equality in the United States.

Making measurements

African people were few and far between in Europe at the time, yet a handful of measurements made of the skulls of African people were used to describe the residents of an entire continent. Samuel von Sömmerring measured the brain cases of an African boy and a 20-year-old man and declared that their skulls were smaller than their European counterparts, and were similar in morphology to the orang-utan. Many other researchers disagreed. In 1836, Freidrich Tiedemann made the point that the people being measured were frequently the most poorly fed and most powerless members of society and should not be seen as typical. The neurologist Paul Broca also attempted to put the measurements on a firmer scientific footing. He believed that the size of the brain was proportional to the level of education, and so assumed Europeans would have the biggest brains.

Political debate

The developments by neuroscientists in North America took into account that there were at least three of the defined races living in large numbers on that continent. Samuel Morton, president of the Academy of Natural Sciences in Philadelphia, had one of the world's largest skull collections. Morton was a polygenist—he believed the populations of each continent were created separately before the time of Noah.

In 1868, Josiah Nott and George Gliddon presented a study that served to reinforce the belief that whites were superior to blacks, and it was natural for the former to dominate the latter group. They drew widely distorted images of African skull anatomy and faces that made then appear closer relatives of apes than Europeans. All of these researchers cherry-picked data to assert, wrongly, that European skulls (and brains) were bigger— and that meant they were more intelligent. The link between brain size and intelligence has never been proven to this day.

DOWN SYNDROME

In 1866, John Langdon Down, a British doctor, made the first description of the genetic disorder that is now known as Down's syndrome. The disorder has many physical and neurological features. Down characterized the total disorder using a racial classification. He called it Mongolism based on the eye shape of people with the disorder. This offensive term is still used in some quarters today.

An 1830s, engraving attempts to catalog the different features of the world's humans, with the western European at the center.

40 Electrical Stimulation

EVER SINCE GALVANI HAD FOUND A LINK BETWEEN MOVEMENTS AND ELECTRICITY, people had been looking for ways to use electricity as a therapy. However, the techniques developed were of more use in research.

For much of the 19th century, neuroscientists used lesions—deliberate cuts—to investigate the functions of the nervous system. This technique worked fine for showing which nerves were related to different body parts, but it was less successful at revealing how the brain itself worked. If anything, cutting the brain appeared not to do much at all, lending weight to the idea that the brain worked as a complex whole and was not divided up as the phrenologists had predicted.

In 1870, two researchers from Berlin published details of a pioneering new approach that used electricity. Eduard Hitzig and Gustav Fritsch began to apply electric currents to the frontal lobes of dogs. They used Hitzig's bedroom as a laboratory and found that by applying current to precise locations on the skull, they could elicit specific movements in the dog. This was proof that movement was controlled, or at least influenced, by the forebrain, overturning older notions that it was entirely controlled by the cerebellum and spinal cord.

Spa treatments were less relaxing in the 19th century than they are today. Looking like a primitive electric chair, this getup was in fact an attempt at using electricity as therapy for general aches and pains.

Photographs from 1862 show Duchenne de Boulogne, a French neurologist demonstrating how facial expressions can be altered with a bit of electricity. Duchenne's work focused on electrifying muscles and nerves, not the brain itself.

41 Mood Disorders

THE YEAR 1870 WAS A BIG ONE IN THE STUDY OF THE BRAIN AND ITS EFFECTS. WHILE THE FUNCTIONAL ANATOMY CONTINUED TO BE INVESTIGATED, an English doctor pioneered a new field of research: Diseases of the mind. Henry Maudsley proposed these were brain disorders expressed through emotion.

"Mental disorders are nervous diseases in which mental symptoms predominate and their entire separation from other nervous diseases has been a sad hindrance to progress."

HENRY MAUDSLEY

BODY AND MIND:

AN INQUIRY INTO THEIR CONNECTION AND MUTUAL INFLUENCE, SPECIALLY IN REFERENCE TO MENTAL DISORDERS.

BEING THE

GULSTONIAN LECTURES FOR 1870,

DELIVERED BEFORE THE ROYAL COLLEGE OF PHYSICIANS.

WITH APPENDIX.

BY

HENRY MAUDSLEY, M. D., LOND.,

FELLOW OF THE ROYAL COLLEGE OF PHYSICIANS;
PROFESSOR OF MEDICAL JURISPRUDENCE IN UNIVERSITY COLLEGE, LONDON;
PRESIDENT-ELECT OF THE MEDICO-PSYCHOLOGICAL ASSOCIATION;
HONORARY MEMBER OF THE MEDICO-PSYCHOLOGICAL SOCIETY OF PARIS,
OF THE IMPERIAL SOCIETY OF PHYSICIANS OF VIENNA, AND OF THE SOCIETY FOR THE PROMOTION OF PSYCHIATRY AND FORENSIC PSYCHOLOGY OF VIENNA.
FORMERLY RESIDENT PHYSICIAN OF THE MANCHESTER ROYAL LUNATIC HOSPITAL, ETC.

NEW YORK:
D. APPLETON AND COMPANY,
90, 92 & 94 GRAND STREET.
1871.

Body and Mind: An Inquiry into Their Connection and Mutual Influence, *the book in which Maudsley set out his most significant psychiatric theory.*

Henry Maudsley's ambition when in medical school was to be a surgeon. However, he became impatient with the slow progress of his career and opted for a stint abroad working for the British colonial authorities in India. To qualify for that role he had to work for six months in an asylum—a hospital for lunatics. "Lunatic" was the term for anyone suffering from a debilitating mental disorder with seemingly no physical problems. It is derived from the ancient belief that mental illness was influenced by the Moon. Maudsley found the work invigorating and switched his career path to psychiatry.

Psychiatry was a largely undeveloped discipline in the mid-1800s, although the Bethlam Royal Hospital in London had been in operation for the best part of six centuries. By Maudsley's time this institution had been renamed the Bedlam Hospital, at first as a nicknamc, and then officially. The word "bedlam," meaning "noisy chaos," is a description of conditions inside.

Mind and Body

Henry Maudsley wanted to work at the Bedlam but failed to win a position. Nevertheless, he had the good fortune to marry Ann, daughter of John Conolly, London's preeminent psychiatrist at the time. With marriage came his father-in-law's private asylum, and Maudsley used it as a test bed for his theories. In 1870, he was ready to present them in a lecture (to the Royal College of Physicians) called Body and Mind. In it, Maudsley proposed that many mental illnesses could be classified according to the emotional symptoms of patients. He termed them "affective disorders," but the term "mood disorder" is more common today.

Maudsley identified three types of mood disorder: Those resulting in depression, mania, or anxiety. Maudsley's contribution was to formalize a medical understanding of how patients used emotional terms, metaphysical ideas, not just physical descriptions, to describe their symptoms. Since his day, mood disorders have been subcategorized in many ways, but the causes of mood disorders are still poorly understood. Mood is now understood in terms of the chemical activity of the brain, but whether that is the primary cause of disorders or merely another symptom is the big question. Among radical theories is that extremes of mood have an evolutionary root, something that Maudsley's friend Charles Darwin would later write about.

42 Nerve Nets

NEUROSCIENCE IS BUILT ON THE "NEURON DOCTRINE," THE MECHANISM BY WHICH NERVE CELLS MAKE UP THE BRAIN. Despite a growing catalog of nerve cells being discovered, it was still a mystery how they worked. The dominant theory was they formed a vast network of vessels.

In the decades after the discovery of the neuron, the limits of microscopy formed a barrier to understanding how they worked. The long, intricate branches of each neuron were clearly visible, but it is difficult to see where one neuron ended and another started. Could that be because the cells were in fact one immense, interconnected whole, a so-called neural reticulum? The answer to this question had far-reaching effects. Supporters hoped this idea would show that the brain could not be subdivided into localized units. Instead it was an holistic organ, all working at once with something akin to the age-old "animal spirits" flowing through the neural transportation system.

Looking for links

Cell theory insisted that any nerve network had to be made of many individual cells that were connected together. It could not be single entity. Researchers devoted themselves to finding the links, or anastomoses, between the cells that would prove the nerve net theory. What was needed was some way of seeing each cell's structure with utmost clarity. Such a technique was not far away.

The neuron doctrine describes the brain in terms of nerve cells forming a vast network. The question was: How did they communicate?

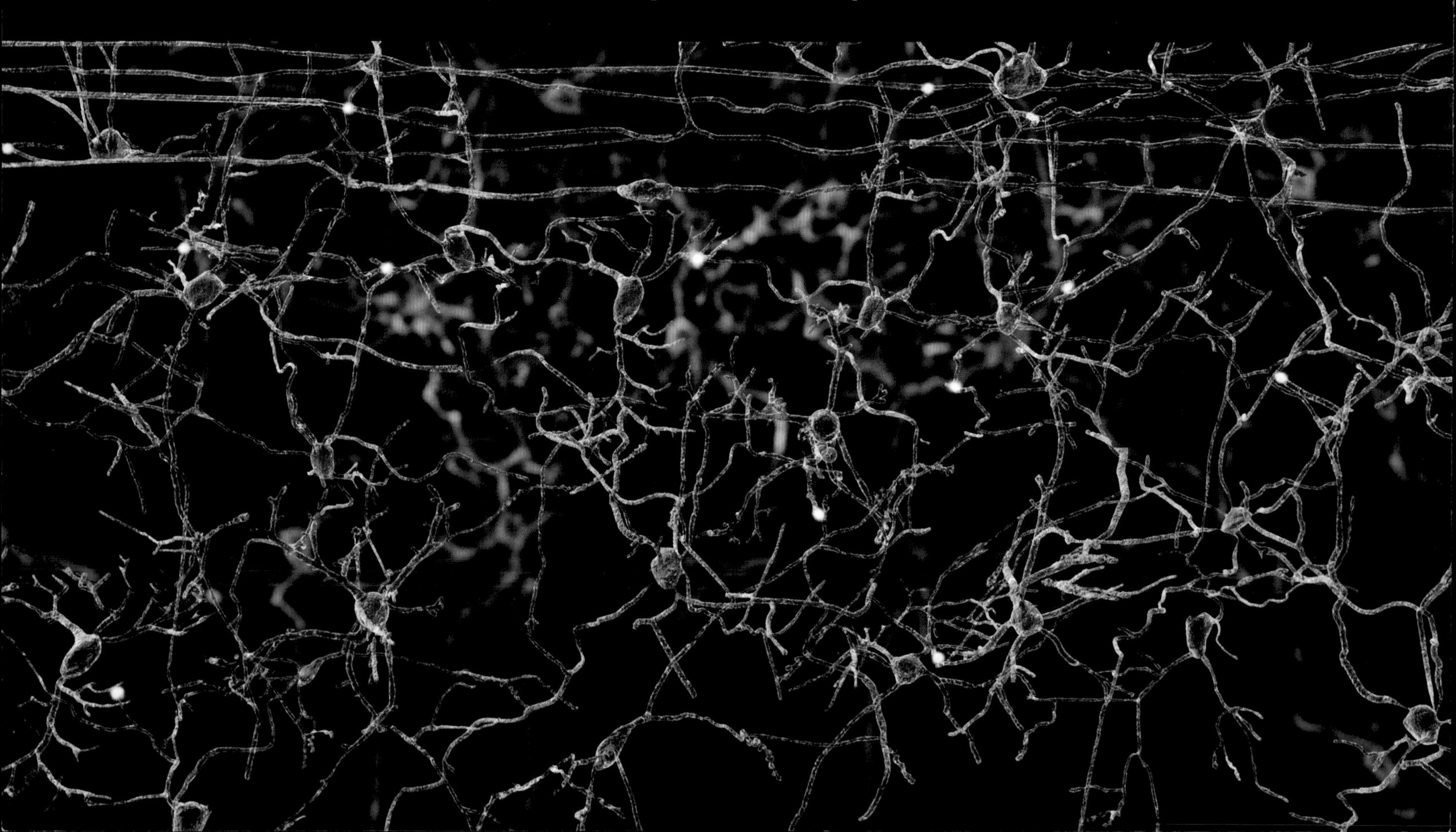

43 Sensory and Motor Centers

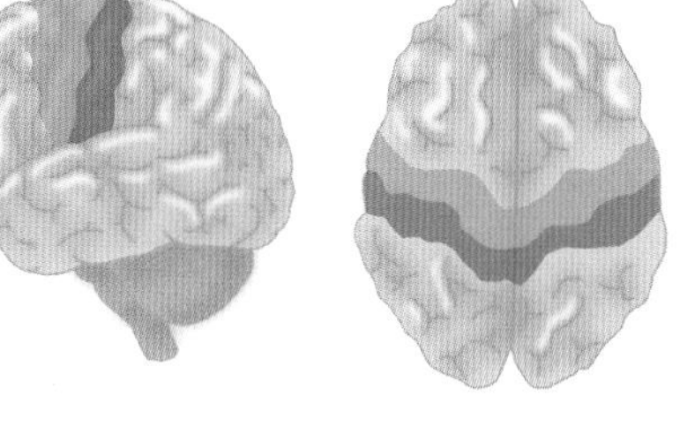

The motor cortex (red) and somatosensory cortex (green) seen from the side and from above in both hemispheres. The motor center is in the frontal lobe, while the sensory one is in the parietal lobe.

ELECTRICAL STIMULATION OF THE BRAIN HAD LOCATED A REGION ASSOCIATED WITH MOVEMENTS. Further experiments confirmed this was the motor cortex, the seat of voluntary body movements. The next question was how did this region relate to the perception of touch and motion?

DAVID FERRIER

Ferrier's work was the first conclusive proof of a motor cortex and paved the way for a better understanding of the touch senses. He used "faradic" alternating currents, which had a stronger effect on muscle action than direct "galvanic" ones. His greatest triumph came when he paraded a hemiplegic monkey before the world's leading neuroscientists in 1886, and was able to show he had caused the condition by cutting away its motor cortex.

Moving research out of the bedroom of Fritsch and Hirtzig, a Scottish researcher called David Ferrier was able to confirm that what the German pair had found was indeed the motor cortex: The region of the brain that handled muscle movements—at least the voluntary ones (see box). Ferrier extended the electrical stimulation work of Fritsch and Hirtzig using monkeys instead of apes. He was also able to isolate the control centers for specific movements like walking and grasping objects.

This was not enough to prove the location of the motor cortex. Ferrier knew this because he could produce similar movements by stimulating several parts of the parietal and temporal lobes. So he began to make lesions in those parts of his monkeys' brains to see what effect they had on movement. After years of experimenting he was able to pinpoint the motor cortex in the precentral gyrus of the frontal lobes. The left lobe controlled the right side of the body, and the right controlled the left side.

Feeling the way

Then followed a disagreement about the role of the motor cortex in sense perception. Some suggested that the reason the muscles stopped moving when the motor cortex was removed was because the brain could not feel anything there. This was shown to be only half true. Lesions in the motor cortex did not stop the skin picking up light touches, although harder pressure was barely perceived at all. Reflexes did still occur (although were often less vigorous). It would take seven decades of research before the somatosensory cortex was mapped just behind the motor cortex in the brain. In the meantime, there were still questions to answer about the way the brain perceived different types of contact—and how it "felt" its own body.

REFLEX ACTION

Not all muscle movements are necessarily controlled by the motor cortex. Many happen seemingly automatically. Involuntary movements like this are known as "reflexes." They are controlled by a simple arc of neurons that runs from the touch receptor (or other similar sense) through the spinal cord and out again via a motor neuron to the muscle. The motor cortex has a veto on these actions, while the sensory cortex makes us fully aware of them.

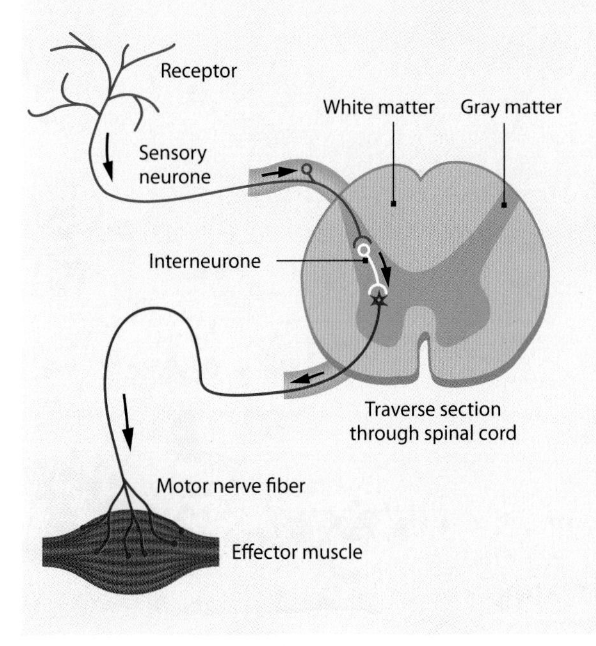

44 Phantom Limbs

WAR PROVIDES AMPLE OPPORTUNITIES FOR NEUROSCIENCE. THE TERRIBLE INJURIES OF THE WOUNDED offer insight into how the brain works. The carnage of the Civil War provided all too many research subjects into the phenomenon of phantom limbs.

Private William Sergeant of Co. E, 53rd Pennsylvania Infantry Regiment, in uniform, after the amputation of both arms.

The earliest record of a phantom limb was made by Ambroise Paré in 1551. Paré, a French barber–surgeon, was a trailblazer in amputations and prosthetics. He also pioneered the use of antiseptics, something that most surgeons would not endorse for more than 300 years to come.

A phantom limb is when a person reports feeling they have a limb even after it has been removed by accident or design. Frequently, a phantom limb hurts. The phenomenon speaks to several areas of research. One is the kinesthetic sense, where the muscles of the limb provide the brain with information about position and posture. This sense is linked to the motor cortex, because damage to that region dulls the ability to feel deep within the paralyzed limb. The cutaneous senses are unaffected by the same damage. They pick up more delicate stimuli from contact with the skin, such as heat and soft pressure. Both are linked to pain, which arises when the stimuli exceed a critical point. It is safe to say that a limb injury that needed amputation falls into this category.

Stump Hospital

The American surgeon Silas Wier Mitchell was stationed at Philadelphia's South Street Hospital during the Civil War. This is where many of the wounded from the Battle of Gettysburg were sent, and the hospital was named Stump Hospital due to the high likelihood that Mitchell and his fellow surgeons would be forced to amputate wounded limbs.

Mitchell made a study of the many men who reported phantom limbs after recovering. In keeping with the supernatural theme, he characterized them as "sensory ghosts," and noted that the phantom limb was frequently shorter than the original real one. Mitchell reported that the hallucination of feeling a limb where there was none could be brought on by the slightest stimulus, such as a cough or a gust of wind. As well as pain, sufferers also felt the arm moving. Wearing a prosthetic often enhanced the sense of having a real limb there.

Phantom limbs are thought to be caused by the severed nerves in the stump continuing to send sensory information to the brain. However, research shows that sufferers are also able to contort their phantom limbs into impossible positions, suggesting the sensations are linked in some way to a persistent mental image of the body's original form.

45 Charles Darwin on Emotions

DARWIN DID NOT REST ON HIS LAURELS AFTER PUBLISHING *ON THE ORIGIN OF SPECIES* IN 1859. He had other interests, too, including pigeons, sexual characteristics, and, of most interest to us, the function of emotions.

Ape expressions do not indicate the same emotions as human ones. In chimps staring wide-eyed indicates a threat, while smiling shows fear.

In 1872, Charles Darwin published *The Expression of the Emotions in Man and Animals*. It was his third book based on his theory of evolution, and like the second book (*The Descent of Man*), it was written because he could not get everything he wanted to say into his first and most famous book.

Darwin was friends with Henry Maudsley. They shared many interests and often discussed the subject of emotion. Darwin's theme was to explore what emotions are for, why humans have them, and how they compare with behaviors that could be characterized as emotional that are seen in other animals, apes chief among them.

Facial expression

Darwin's focus was on how emotion is communicated through facial expressions, something we share with apes (and other animals). He proposed that these expressions arose from a basic physical purpose. For example, snarling shows the teeth as a warning. The emotional response that a person felt—anger, sadness, etc.—was a kind of mental reflex, allowing the body to respond appropriately to a situation without having to think too much about it. Inappropriate emotion, such as rage or paranoia, Darwin suggested, were due to an over activation of the nervous system.

Darwin illustrated his points with pictures of actors performing various emotional states. Seen here: Pride (one of which is literally hair-raising) and resignation.

46 The Structure of the Eye

THE NEXT SENSE ORGAN TO FALL BEFORE THE MIGHT OF THE MICROSCOPE WAS THE EYE. Successive anatomical breakthroughs accrued through the 18th and 19th centuries. However, to figure out how the eye worked, a bit of chemistry was also needed.

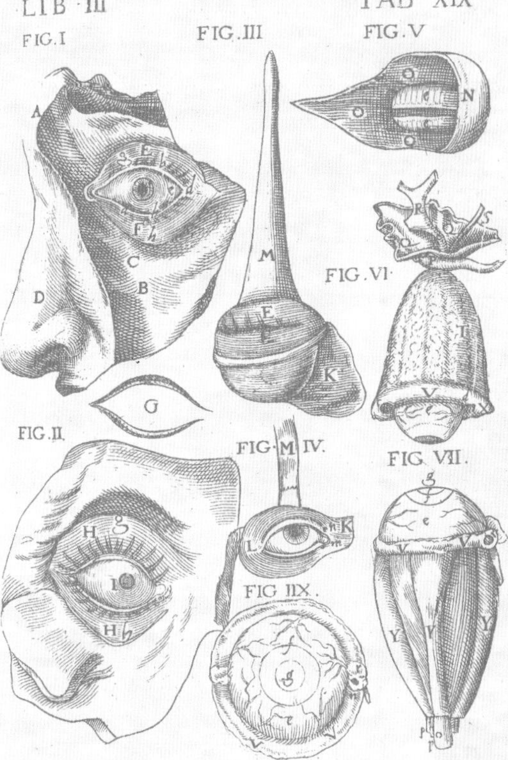

A 17th-century anatomy of the human eye, its orbit, and musculature drawn in 1605 by Gaspard Bauhin, an avid anatomist.

One of the drawings Maximillian Shultz made of the retina cells in 1872.

Much of the anatomy of the mammalian eye was known prior to the advent of the microscope, mostly thanks to Galen. The Roman doctor described the conjunctiva, which forms the outer membrane at the front of the eye. Under that is the cornea, which covers the colored iris and dark pupil, familiar to all of us that have stared into another's eyes for whatever reason. Then comes the lens, the window into the main body of the eye, which is filled with a clear gel: the vitreous humor. The white sheath around the eyeball is the sclera (very visible in we humans and thought to help make our eyes more expressive). The next layer in is the choroid, which forms the foundation for the retina. Now understood as the light-sensitive part of the eye, Galen and his colleagues could do little more than describe its appearance, which reminded them of a fishing net. (The word retina means "like a net.")

Alhazen had shown that the eye worked like a camera obscura, with the pupil letting light into the "dark room" that was the eye. The crystalline lens focused the light onto the back wall, the retina. In 1668, Edme Mariotte had discovered a blind spot where the optic nerve penetrated the retina. The next step revealed that visual information from parts of each eye seemed to be compartmentalized at the optic chiasm. The process of how light influenced the nerves was a mystery.

Receptors and pigments

In 1791, Samuel von Sömmerring reported seeing the blind spot as a shallow dip at the center of the retina. The blind spot is actually off-center. What Sömmering had seen was the fovea,

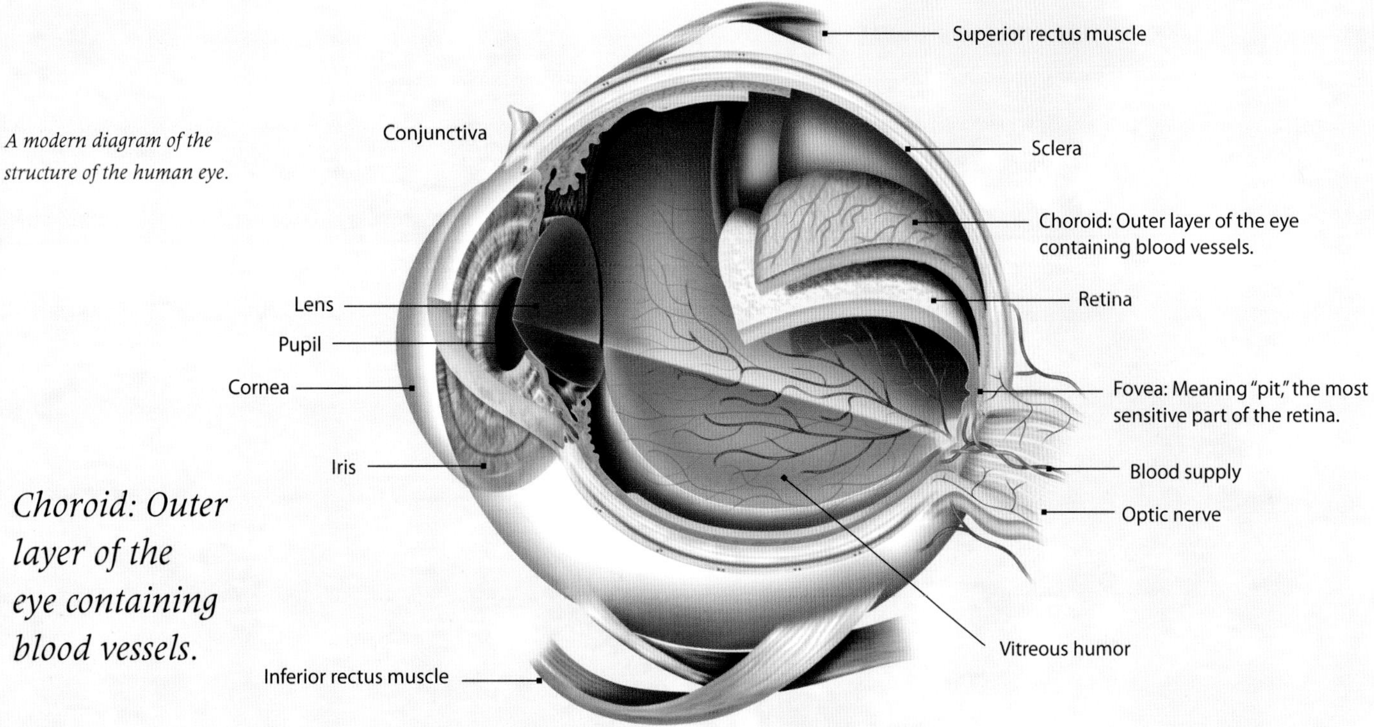

A modern diagram of the structure of the human eye.

Choroid: Outer layer of the eye containing blood vessels.

Fovea: meaning "pit," the most sensitive part of the retina.

the most sensitive portion of the retina. Early looks at the retina through a microscope revealed what one researcher saw as rods, or cylindrical papillae, which were assumed to be the light-sensitive receptors. It was found that each receptor had an individual connection to the optic nerve via some supporting retinal cells.

The big breakthrough was made by Franz Boll in 1876. He saw that a detached retina changed color, bleaching to yellow from red-purple, when held to the light. Boll's color-changing pigment has since been named "rhodopsin." Rhodopsin's bleaching is caused by energy transferred from absorbed light. Cone-shaped cells, which dominate the fovea, were found to contain three other pigments, each adapted to absorb light of a certain color. Rod cells, therefore, are used for monochromatic vision in low light levels, while cones produce sharper color vision during the day.

THE RETINA

When the structure of the retina was seen in detail, it appeared to be upside down. The photosensitive cells are buried under two other layers. Light has passed through the ganglion cells and bipolar cells to reach them. The bipolar cells connect to several rods or cones (not both) and collect their responses to light. These signals are then sent to the ganglion cells which connect to the optic nerve.

Light arrives
Ganglion cells
Bipolar cells
Rods and cones
Pigment cells

47 The Black Reaction

THE FINE DETAILS OF NEURON STRUCTURE, ESPECIALLY HOW THEY CONNECTED TO ONE ANOTHER, CONTINUED TO ELUDE RESEARCHERS. Many scientists raced to make their names by finding techniques of isolating single nerve cells among the jumbled mass of brain tissue.

Sigmund Freud is famed today as the founder of psychoanalysis but he had an earlier attempt at scientific fame. He had harbored an interest in the brain in general and neurons in particular while still in medical school and researched the nervous systems of crustaceans and cephalopods (which have particularly large axons). Within a few months of beginning to practice in a Vienna hospital in 1883, Freud had decided to specialize in neurology, and set about looking for a new staining technique that would revolutionize microscopy and reveal the true nature of nerve cells. However, he was already ten years too late—only no one realized it.

A neuron stained using the Golgi method, showing the cell's individual branches.

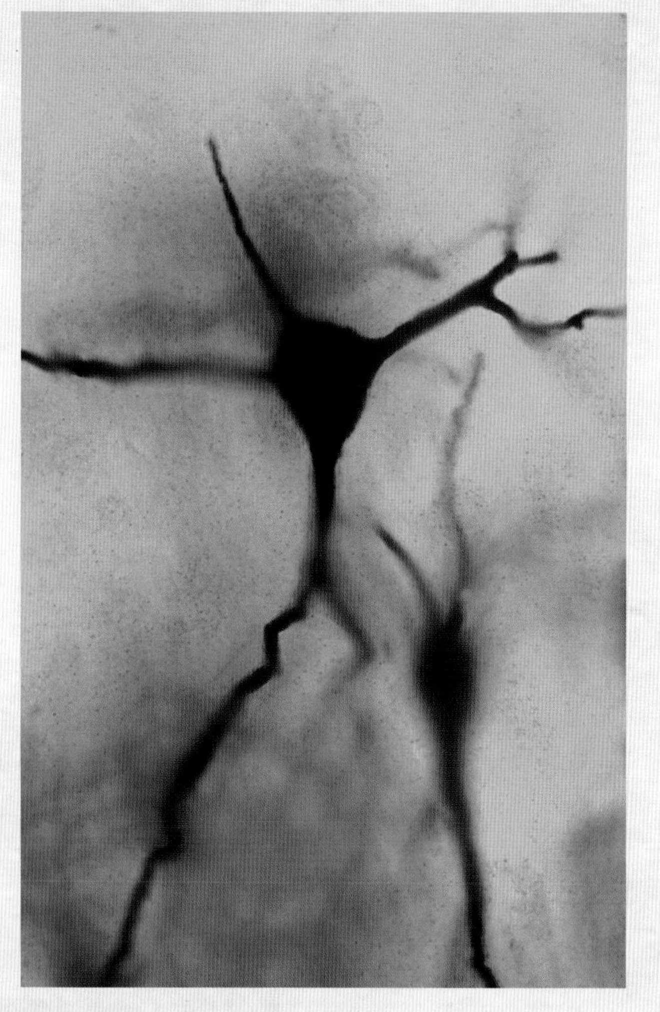

"I am delighted that I have found a new reaction to demonstrate even to the blind the structure of the interstitial stroma of the cerebral cortex."

CAMILLO GOLGI

Kitchen researcher

In 1872, the Italian Camillo Golgi physician began a residency at a hospital for incurables near Milan. In his spare time he repaired to the hospital kitchen, not to eat but to cook up a new staining technique. A common preparation was to harden specimens of brain tissue in chromic acid, and Golgi tried to do the same thing with potassium dichromate, a related salt. He then added silver nitrate, the same chemical used in photographic paper, to the specimen. It appears that the ensuing results were achieved by good luck, but they were a great success. In 1873, Golgi announced his discovery, which he called *la reazione nera*, or "the black reaction."

For reasons still unknown, Golgi's preparation only stained about five percent of the cells in the specimen. The others remained invisible in the yellow background, but the stained cells stood out black as individual units. Golgi found that the tiny internal structures, or organelles, of the cells were also visible. One common organelle is named the Golgi apparatus in his honor. Golgi hoped that this technique would finally reveal how neurons connected. However, it would take another 25 years, and he would not like the result.

48 Intentionality

BY 1874, THE PHYSICAL AND MENTAL WORLDS HAD BEEN DESCRIBED IN GREAT DETAIL, but they were still regarded as being made of substances. Drawing a philosophical distinction between the two is something that would help inform the understanding consciousness.

Physical substance has heft and form, while mental substance is weightless and diffuse, but unless stated otherwise both must be made of the same "thing."

German philosopher Franz Brentano swept away the need for this interpretation. What differentiated the worlds was "intentionality," he said. The mental world had it—ideas always refer to something—while the physical world did not. It just is. This block of paper, board, ink, and glue in your hands has no intentionality. However, when you read it, the ideas in your head do. To Brentano, even something as visceral and "thoughtless" as pain had intentionality—to warn you about damage to the body.

Brentano sought to clarify the mental world from the physical one.

"An unconscious consciousness is no more a contradiction in terms than an unseen case of seeing."

FRANZ BRENTANO

49 The Microtome

THE 1875 INVENTION OF A MICROTOME, a machine for slicing the brain, revolutionized the study of the brain's anatomy and with it neuroscience.

The microtome, or thin section cutter, was invented by Bernhard von Gudden, the personal physician to the "Mad King" of Bavaria, Ludwig II. Von Gudden's device could slice the brain into thousands of sections, making it possible to scrutinize the cellular structure of the brain in more detail than ever. Before its invention, researchers used their own preferred techniques to produce specimens for study. These ranged from scraping off tissue with a knife or peeling it away with tweezers. The lack of precision led to widely varying results, and meant it was impossible to repeat and verify dissections.

The Gudden microtome changed all that with a technique called "serial sectioning." The many thin sections made it possible to turn the three-dimensional brain into a series of two-dimensional "snapshots," which, when viewed in order, made a very precise representation of the entire brain. Also, the thin sections were ideal for use with microscopes, giving neuroscientists the clearest view yet of brain cells.

Bernhard von Gudden's design of a microtome could cut slices of brain that were just two thousandths of an inch thick.

50 Electrical Encephalography

IN 1875, IT HAD BEEN DISCOVERED THAT THE BRAIN, IN COMMON WITH THE REST OF THE BODY, HAD AN ELECTRICAL SIGNATURE. Within 60 years, the electrical activity of the brain would be used as a diagnostic and research tool into the cerebral cortex, the outermost layer of the brain.

"We see in the electro-encephalogram a concomitant phenomenon of the continuous nerve processes which take place in the brain."

HANS BERGER

It would sound trite to say that Hans Berger's invention of the electroencephalograph in 1924 was the result of a brain wave. Instead, it all began with a near-death experience. However, we are getting ahead of ourselves. In 1875, an English doctor called Richard Caton discovered that he could detect a weak electrical field produced by the exposed brains of live rabbits and monkeys. He published the results, but they were ignored. In 1890, a Pole called Adolf Beck found that the electrical signature of animal brains—detected by inserting electrodes directly into them—could be altered by stimulating them in different ways. Recording over a long period, Beck began to see what appeared to be rhythmic oscillations—the first brain waves. Other researchers dabbled in the field, and in 1903 the Dutch scientist Willem Einthoven developed the electrocardiograph (ECG). This records the electrical activity of the heart muscles. The electroencephalograph (EEG), created by the German Hans Berger, would do something similar, but to begin with he had other ideas.

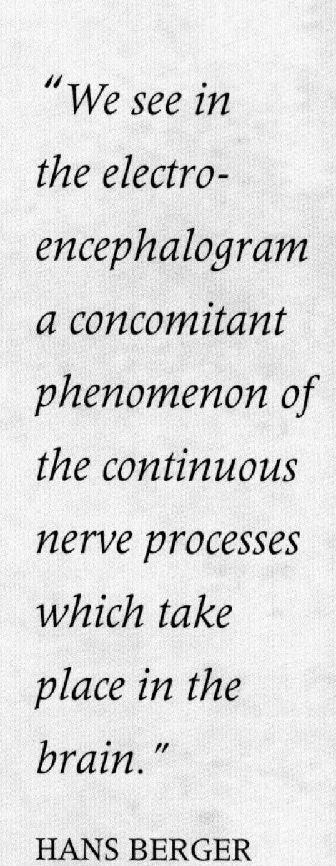

Hans Burger, inventor of electroencephalography (EEG), and his early EEG trace showing an alpha brain wave. EEGs are also used by doctors to diagnose epilepsy and other seizures. During these episodes, the brain activity shown on the EEG becomes very jumbled.

Psychic research

As a child, Berger's chosen career was to be an astronomer, but after one semester of college he dropped out and joined the army. While on maneuvers he was nearly crushed under the wheels of a thunderous cannon, but escaped unhurt. His sister reported feeling a pang of fear at the same moment as Hans's brush with death, and told him by telegram. Berger's life was changed by the apparent telepathy. He vowed to find out where the "psychic energy" was produced in the brain.

The result was the first functioning EEG that produced a record of human brain activity. Berger called his device a "brain mirror," and his lack of understanding of how it actually worked led many psychiatrists and neuroscientists to discount the machine. It did not help that popular descriptions of the device pronounced that its jagged graphs were a unique representation of a person's personality, a fingerprint of the soul. Nevertheless, Berger persisted and his device was gradually accepted. Today, every hospital in the world has one of his brain mirrors.

Brain waves

A look in the brain mirror reveals that when a person is relaxed, resting with their eyes closed their brain's electric field oscillates in a smooth pattern, rising and falling 20 times a second. These are alpha waves. When the person is thinking actively, the brain produces beta waves, which are more erratic and spiky. Long, slow delta waves are associated with sleep, while gamma waves are thought to play a role in conscious awareness. However, as doctors wondered at the potential of Berger's machine, the role of electricity in the brain was still something of a mystery.

51 Hypnotism

TODAY, HYPNOTISM HAS FOUND ITSELF A PLACE IN ENTERTAINMENT AND IN THERAPEUTIC APPLICATIONS. However, to get there it has had to shed its reputation as an eerie force that preys on weak minds.

Hypnotism is a state of awareness where a subject becomes highly focused on one thing or person and is less aware of what else is happening around them. There is disagreement about whether the brain enters an altered state, or whether the subjects merely become willing to be guided by another's suggestions.

In the late 1870s, the French neurologist Jean-Martin Charcot investigated the link between hypnosis and hysteria (a term then used to describe a range of emotional disorders, especially in women). Charcot regarded hysteria in a similar light to epilepsy, in that sufferers had a nervous system predisposed to it. He proposed that the same predisposition meant they could be hypnotized. This idea was soon discounted, and the term hysteria has since been removed from the medical lexicon. However, a result of the research was that people began to distrust hypnosis, seeing it as a means to force someone to act against their will.

Jean-Martin Charcot gives one of his many demonstrations of hypnosis with a "hysterical" patient. Doctors and the public in general began to suspect that certain people could be hypnotized to commit crimes, which led to a debate on criminal responsibility and mental illness.

52 Narcolepsy

IN 1880, A NEW TERM WAS INVENTED TO DESCRIBE AN UNUSUAL CONDITION WHERE PEOPLE APPEARED TO GO TO SLEEP FOR SHORT PERIODS. The word "narcolepsy" means something like "seized by sleep" and was coined to differentiate this condition from other seizures and trancelike states.

The word *hypnosis* means to "make sleep," and refers to the way hypnotized subjects were thought to be entering a state between wakefulness and sleeping. Narcolepsy may seem superficially similar, but in 1880 Frenchman Edouard Gélineau sought to define it more fully. He took as his subject a 36-year-old merchant, who suffered 200 "sleep attacks" each day. The attacks began with a feeling of sleepiness, and were followed by five minutes of deep sleep, where the body was totally relaxed and unmoving. Sufferers reported dreaming vividly during attacks but seldom remembered much of the content of the dreams.

The attacks were most frequent during periods of either calm or elevated emotion. Physical activity tended to reduce the attacks. Narcolepsy was not the product of sleep deprivation, although the symptoms were similar. The hunch was that the condition runs in families, and this has since been confirmed. One suggestion was that it is a vestigial behavior from distant ancestors who would "play dead" to avoid predators.

53 The Visual Cortex

THE 18TH-CENTURY INFORMATION SENT FROM THE EYES WAS TRACKED AS FAR AS THE OPTIC CHIASM. From there the optic tract connected to the center of the brain. Was this where the brain conjured visual perception? The hunt was on.

"Upon gently pressing the dura mater, he suddenly perceived a thousand sparks before his eyes."

HERMAN BOERHAAVE

With the benefit of hindsight, we look back on phrenology with a humoring gaze. How could anyone think the shape of the head indicated a person's character? Phrenology's chief proponent, Franz Joseph Gall, located visual skills around the eye, as one might expect. We now know that it is the back of the head that plays the leading role in vision. And one of the reasons we know that is thanks to some first-class anatomical research by none other than Franz Joseph Gall!

After their crossing point at the optic chiasm, each optic tract continues on to the lateral geniculate nucleus, part of the thalamus. Gall and his associate Johann Spurzheim found that this region began to atrophy, shrink, and fade away, when the

ACQUIRED COLOR BLINDNESS

Regular color blindness is caused by one or more types of the color-sensitive cone cells not working in the eye. However, strokes or other brain insults could lead to the loss of color vision. One of the first recorded cases was in 1882. A stroke sufferer could see perfectly well but was unable to discern colors. An autopsy of the brain revealed damage to the lingual gryus, close to the primary visual cortex.

optic nerves have been severed, as did the superior colliculus, a bulging body in the midbrain. These sites were deemed the end points of the optic nerve. Lesion studies of these areas showed they were crucial in eye function. The nerves that control the movements of the eye emanated from the midbrain close to the superior colliculus. However, it was believed that projections of the images captured by the eyes, were being sent on to the cerebrum.

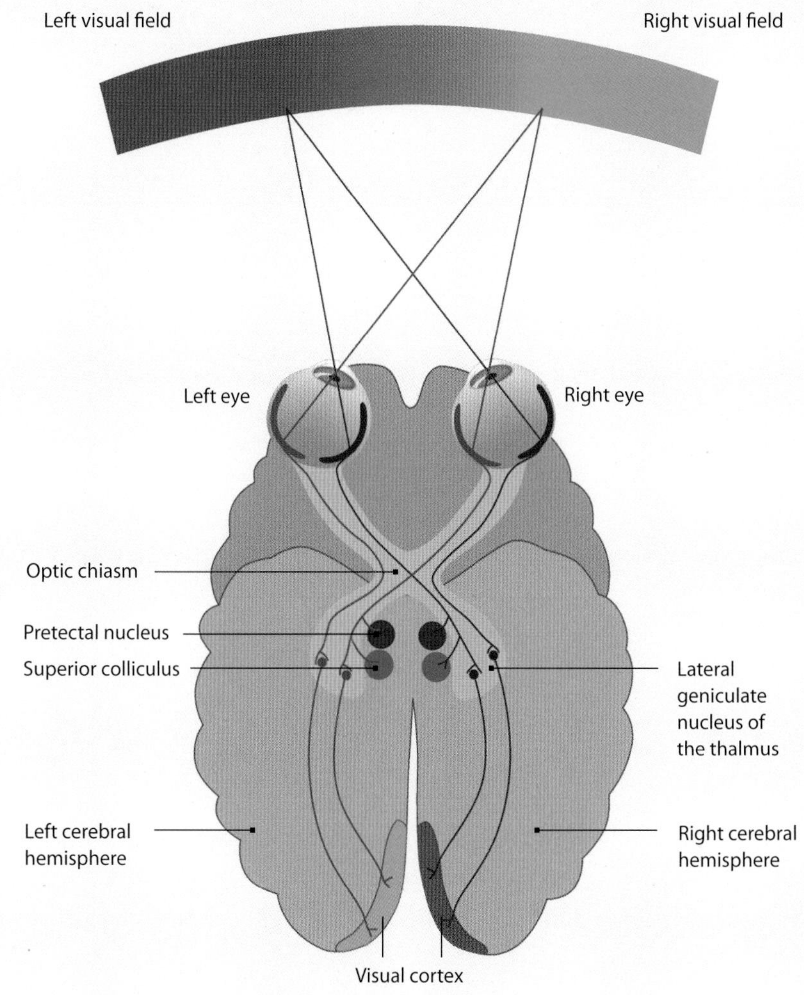

A diagram of the visual pathway through the human brain. It shows how objects to the right of the body are perceived in the left hemisphere, and vice versa with objects on the left side.

Back of the head

The Dutch scientist Herman Boerhaave had reported the first evidence that vision was managed in the cerebral cortex. He met a beggar in Paris in the 1730s, who collected money—rather successfully, no doubt—in a dish made from a piece of his own skull that had been removed years before. For the right price, people were allowed to press his exposed brain. Boerhaave reports that a gentle press led to visual sensations. He pressed harder, making the man blind, and eventually fall limp, and unconscious! Boerhaave then waited to see if the beggar recovered. He did, with his sense of sight being last to return. In 1776, Francesco Gennari described a stripe of pale tissue running through the occipital lobe. This feature, the Line of Gennari or the striate cortex, would later be found to extend all the way to the lateral geniculate nucleus, and was formed of a multitude of axons carrying signals to the back of the brain.

Primary visual cortex

By the 1870s, reports were coming in that damage to the occipital lobe on one hemisphere led to blindness in the visual field on the opposite side. The Line of Gennari was the primary visual cortex, but vision was also influenced by other parts of the occipital lobe and adjoining parts of the temporal lobe, too. Hermann Munk is credited with being the leading figure in this discovery. In 1881, he made lesions in the upper occipital lobe of dogs, and nursed them back to full health. Even after years of recovery, Munk found that although the dogs could see perfectly well, their brain damage stopped them recognizing once-familiar objects. This suggested that the visual cortex stores shapes and image concepts that are used to make sense of vision.

54 Tourette's Syndrome

NAMED FOR THE PIONEERING NEUROLOGIST, GEORGES GILLES DE LA TOURETTE, THIS DISORDER IS A COMPLEX MIX OF involuntary movements and vocal tics and cursing.

"Coprolalia: the use of obscenities"

"Echolalia: repeating the words of others"

Tourette's syndrome won its name in 1884, when the young French doctor published his seminal paper on it. However, the disease can be traced farther back to Englishman Samuel Johnson, an 18th-century man of letters famed for writing the first dictionary of English. Johnson was an avid conversationalist and is oft quoted even today, yet he was reported to whistle and moan between phrases and wave his arms around erratically. Seen then as eccentricity, today it sounds like the symptoms of Tourette's.

The first medical account of the disease was made by American George Beard who recorded details of the "jumping Frenchmen of Maine," a community of foresters, many of whom displayed similar symptoms. They were "echolalic," repeating the words said by others and "echopraxic," imitating each other's actions. The "jumping" referred to sudden, impulsive movements they made, often accompanied by a bark-like shout.

Tourette was able to link this affliction with similar reports from around the world. In Malaysia, "jumpers" were said to suffer latah; in Siberia it was known as miryachit. He recorded that men were more likely to suffer than women, and it affected people of all walks of life, education and moral standing. He added to the list of symptoms, including facial tics and "coprolalia," where sufferers swore involuntarily, not at random but frequently in the context of who they were with. The cause of the loss of controls over speech and movements is still unknown. However, the disease is thought to be primarily genetic and has been expanded to encompass a range of "tourettisms."

Georges Gilles de la Tourette wrote about the disease that bears his name soon after starting his medical career.

The jumping Frenchmen of Maine, a community in which Tourette's syndrome was first identified. George Beard reported: "It was dangerous to startle them in any way when they had an axe or knife in their hand."

55 The James-Lange Theory of Emotion

As well as being a pioneer of psychology, William James was an influential philosopher.

IN THE MID-1880S, TWO PSYCHOLOGISTS WORKING INDEPENDENTLY BOTH CAME UP WITH A NEW WAY OF UNDERSTANDING EMOTION. Their interests lay in which came first, the physical or mental response.

As psychologists, the American William James and Dane Carl Lange studied the mind, motivations, and behaviors by breaking down mental processes into their simplest components. When it came to understanding emotions they imagined the following kind of scenario: A large, snarling dog is running toward you. What happens next? Your muscles tense up, your heart starts to race, the blood drains from the face, your stomach sinks. You are ready to run—fast. You will also be aware of an emotion: A fear, which you will associate with all those physical changes, plus there will be a mental tightening of your awareness. You have one thought—to get away from the dog.

How does this make you feel? Frightened? Are you getting ready to run because you feel so scared, or do you feel frightened because your body is getting ready to run all by itself?

James and Lange were not convinced that the physical arousal of the body was the result of the mental awareness, which was the common sense view of everyone from Avicenna to Charles Darwin: The conscious mind's appraisal of the situation stimulated an appropriate physical response. James and Lange believed (they had no way of proving it) that the sensory apparatus of the brain sent commands to the body, which fed back into the conscious mind. That meant the emotional response was a secondary effect. The body did what was required, and the mind just played along.

Opposing views

Not everyone agreed. Some pointed out that the theory meant paralyzed people would never feel emotion. A modern understanding is that physical and emotional responses are not connected but may feed off each other, heightening the effects.

56 Cerebral Dominance

THE DISCOVERY THAT SOME PARTS OF THE BRAIN WERE SPECIALIZED TO CONTROL CERTAIN FUNCTIONS WAS A GREAT TRIUMPH FOR NEUROSCIENCE, but also a cause for concern for some. If different regions had different jobs, which bit was in charge? Was it the moral, thinking human brain, or the animal lurking within?

JEKYLL AND HYDE

Robert Louis Stevenson's 1886 book tells the story of Dr Jekyll, a well-liked and respectable man of medicine, who formulates a potion that allows him to become Mr Hyde, a hideous monster of a man. Mr Hyde is not held back by the same morality as Dr Jekyll, and through him Jekyll is able to indulge his animal urges—that is unless he is found out...

A poster for a 1931 movie of the classic horror story resembles a prequel to The Incredible Hulk.

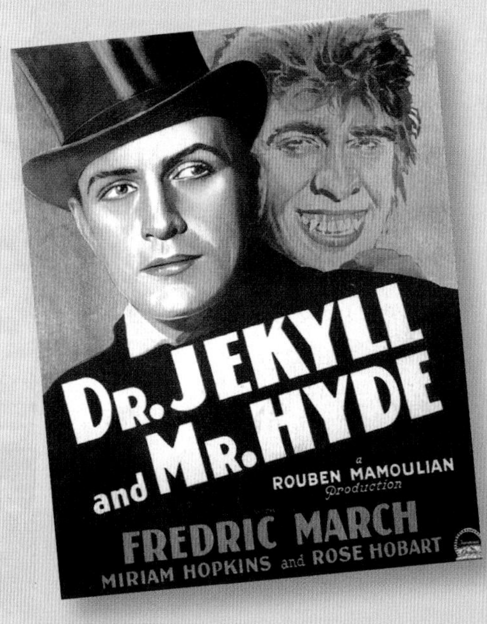

In 1886, Robert Louis Stevenson, the already famed author of *Treasure Island*, published *The Strange Case of Dr Jekyll and Mr Hyde*. This is the story of one person who took on two identities, one a moral, sociable man, the other a murderous ruffian. The book was perfectly timed in the way that it personified a growing unease that the human personality was under the control of competing regions of the brain. If the self-controlling, moral faculties of our brain were disabled, would we all become wild animals?

Two halves of a whole

That concern in the public imagination was the result of a long story that had been developing in neuroscience over the structure and function of the brain.

Since antiquity it had been assumed that the brain's two cerebral hemispheres were identical mirror images of each other. By the 1800s, people were suggesting that mental illness was a loss of unity between the two halves. Injuries to one side of the brain were seen as throwing the two hemispheres out of balance. A French scientist, François–Xavier Bichat, even suggested the way to solve these problems was to hit the healthy side of the head to balance things out a bit.

The left and right side of the brain are not the same: The left is more concerned with words and logic, while the right is devoted to space, emotion, and aesthetics.

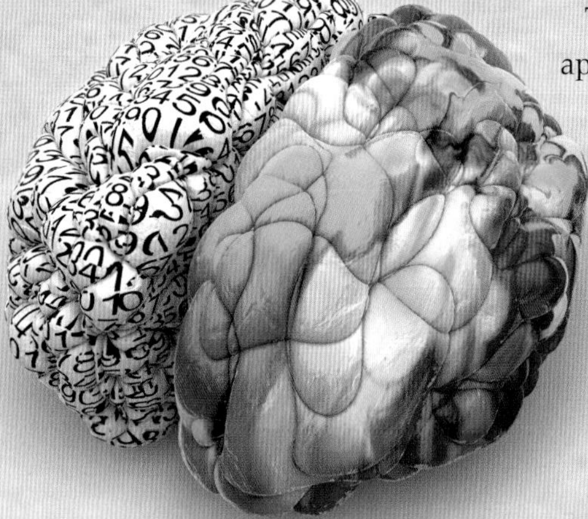

The well-publicized story of Phineas Gage, and his apparent (and oft exaggerated) descent from mild-mannered man to raging brute, fueled the idea that a primitive, animal brain lurked within everyone, and had to be kept under control by the higher, human parts.

Taking sides

Paul Broca's discovery of the speech center not only proved that brain function was localized, it also showed that the left and right sides of the brain were not the same. Broca himself initially resisted the idea, but analysis of stroke and other brain-injured patients suffering from aphasia—the loss of speech—showed that the damage was always on the left hemisphere. In 1874, the German Carl Wernicke also found another region of the brain associated with a form of aphasia. Wernicke's area is also on the left hemisphere, but on the temporal lobe. Damage to Broca's area creates problems in saying words in a fluent intelligible way; damage to Wernicke's area did not hinder speech, but it just did not make any sense.

Broca suggested that the hemispheres were linked to handedness, saying that most people are right-handed, but a left-handed person had speech areas on their right hemisphere. (About a fifth of lefties have this feature.) Broca introduced the idea that one side of the brain was dominant over the other. His suggestion was that the left hemisphere grew faster than the right, and took over in the womb. Broca believed that the most intelligent brains showed the most asymmetry. (This elaborate concept of cerebral dominance has been discredited.)

"I thus drew nearer to the truth... that man is not truly one, but truly two..."

DR HENRY JEKYLL

Word and feelings

Around the same time, the English researcher John Hughlings Jackson showed that damage to the right hemisphere caused problems with spatial awareness, in stark contrast to similar lesions over on the left. He also discovered that victims of aphasia were still able to say platitudes and swearwords, which indicated that these were emotional vocalizations that were emanating from the right brain.

It was becoming increasingly clear that the brain's functions were highly localized. But to understand the workings of brain, it is not essential to follow the idea that one side rules the other.

Nineteenth-century neuroscientists studied the heads of criminals to figure out if their immorality was due to an overly large right side of the brain.

BRAIN TRAINING

If the ideal human had a brain ruled by his or her left side, did that mean less than ideal people had brains that worked the other way around? Would that mean murderers and thieves were slaves to their inferior, uncivilized, impulsive right brains? This led to attempts to educate and "civilize" the right brain, focusing on language skills and physical exercises involving only the left hand.

57 Psychoanalysis

DESPITE A LONG SEARCH AND MANY FALSE DAWNS, AS THE END OF THE 19TH CENTURY ROLLED INTO VIEW, THE IDEA WAS DEVELOPING THAT some mentally ill people had perfectly healthy brains. The only explanation was that these people's afflictions were due to the content of their minds, perhaps without them even knowing it.

Joseph Breuer, a colleague of Sigmund Freud, developed the talking cure method that was eventually popularized by Freud.

The first psychoanalyst was Sigmund Freud, but the field really began with others. Freud, an Austrian and trained medical doctor, began his career as a neurologist. His first field of research was the use of cocaine as a stimulant for his patients—a "project" he continued to dabble in for many years to come. (Cocaine had been recently isolated and was entirely legal, its damaging effects as yet unknown.) In 1885, Freud went to work under Jean-Martin Charcot in Paris's Salpêtrière Hospital. Charcot was an advocate of the "dynamic lesion," a nonphysical problem that damaged the brain temporarily. The lessons Freud took back to Vienna were firstly that mental illness could be due to the impact of the mind and not the physical nature of the brain.

FREUDIANISM

Many of Freud's phrases have entered informal usage, including ego, penis envy, pleasure principle, love-hate relationship, anally retentive (meaning uptight and controlling). The Freudian slip is where our subconscious feelings, often about sex and violence, leak out as we mis-speak about a related subject.

The Greek myth of Oedipus, who killed his father and married his mother, is central to Freud's theories.

Talking cure

The following year, Freud set up a private practice and began to see patients. He used a technique that had been developed a few years before by his friend Joseph Breuer, who encouraged his clients to talk about their feelings while under hypnotism. The idea was that the hypnotism removed inhibitions. Breuer's first success was with a patient anonymized to Anna O. She reported a great success and termed the technique the "talking cure." The name has stuck. Freud then began to formulate the theories that would make him famous the world over.

Unconscious feelings

Freud contended that the source of his patients' mental illnesses, often things like depression, anxiety, and paranoia, stemmed from disturbing desire or memory. These were too disturbing for the mind to admit, and had been locked away in the unconscious. However, they were also too powerful to stay hidden, and leaked out in incongruous ways, causing illness. Freud tweaked the talking cure to tease the pathological thoughts out of hiding. Once confronted they would cease to cause a problem, he thought. This concept of purifying the mind, or catharsis, is a powerful one that has found its way into the modern idea of "closure." Freud's main system was using free association, where patients blurted out the first thing that came into their heads in response to verbal and visual cues, helping to reveal their true feelings.

Theory of mind

Freud believed that even healthy minds operated in the same way. We all repress our darkest desires in the subconscious "id." These interact with our rational mind, or psyche, creating the "ego." The ego is our sense of self, the pilot's seat for our minds. However, there is also a "superego," a dimly imagined mission control that steps in to override the ego on occasion, and is the main cause of mental illness. Freud theorized that the repressed contents of the id were the product of our earliest desires and the relationship with our parents: A boy resents the bond between his parents and wants to kill his father and marry his mother. Meanwhile, a girl feels she has been castrated at birth and so hates her mother and her lifelong "penis envy" drives her to possess men and children. If you think this idea is shocking, Freud would say it is because your ego is protecting you from the truth. Others might look to Freud's own atypical childhood for the origins of the ideas.

Sigmund Freud, a towering figure in the field of clinical psychology and psychiatry.

CARL JUNG

Carl Jung was a contemporary of Sigmund Freud. However, the pair did not have a meeting of minds—or of unconsciousness. Jung (left) thought there was more to the unconscious than Freud's view. Deeper below our personal id, he said, there lay a collective unconsciousness populated by archetypes. These were the fundamental human characters such as mother, father, teacher, hero, villain, along with pivotal events: Death, birth, and "motifs": creation, apocalypse ... Our unique selves are the product of these contradictory unconscious entities being reconciled in the mind. Mental illness is the result of a "complex" of unconscious cues that focus on a particular problem.

.G. Jung.

58 Sleep Deprivation

THE STUDY OF SLEEP IS COMPLICATED—THE SUBJECT IS ASLEEP MOST OF THE TIME. EARLY ATTEMPTS AT RESEARCH IN THE 1860S HINTED THAT sleep follows a cycle of raised and lowered awareness. In 1890, a new approach was made by seeing what happened if a person did not sleep at all.

Being prevented from sleeping is torture—literally. It has even been suggested that the most heinous punishment handed down by ancient Chinese judges was execution by sleep deprivation (it took a long time). In the 1890s, several researchers began controlled experiments on the effects of sleep deprivation. Healthy men were kept awake for 90 hours. The first night passed without incident; by the second night, the subjects would drop off when sitting. Reaction times slowed considerably, movements were slow and imprecise, and the men's memories failed them. By the third night, at least one of the subject was hallucinating, seeing a swarm of colored dots around him, which he attempted to swat away. All three men recovered after a long sleep. However, they did not need the equivalent of three-nights-worth of sleep to get back to normal.

59 Whole Brain Function

WHILE SOME NEUROSCIENTISTS FELT SURE THAT THE BRAIN WAS DIVIDED INTO FUNCTIONAL ZONES, many others refused to accept the idea. Where did one zone end and another start, they wanted to know. In their eyes, the brain was built to work as a whole.

The school of neuroscience that subscribed to the "whole brain" view was known as "holism." Followers included Camillo Golgi, whose staining techniques were showing up the cell structure of the brain in greater detail than ever before. Golgi firmly believed this work would reveal physical connections between the nerve cells. Some holists thought that cells were fused by minute channels linking the dendrites of one cell with the axons of another. Others, like Golgi, believed that all connections were made between axons. The fundamental reason behind his theory was the velocity of thought and nervous responses. Only a flow of material running directly between cells could account for the sheer speed at which the brain worked.

An inkblot speaks to each of us in different ways as we draw on a variety of faculties in the brain to give it meaning.

Decorticating dogs

The most vocal supporter of holism was Friedrich Goltz. The German publicly opposed the findings of Ferrier and others who had proposed locations for muscle and sensory control in the parietal lobes. Goltz cut away large chunks of dogs' brains, in a process he called "decortication," to show that they were still able to move and sense the world. They were, however, "idiotic," in his words, because the reduction in the overall brain made them less intelligent.

Hermann Munk, who was instrumental in locating the visual cortex, took a half-way position: The location of sensory and motor functions could be pinpointed, but higher "executive" functions could not. Jacques Loeb pointed to the brain's ability to recover from injury as evidence that the brain worked by "association processes happening everywhere in the hemispheres ... not as a mosaic of independent parts."

From top to bottom: Goltz, Munk, and Loeb, three champions of holism.

60 Touch Sensors

BY 1894, MOST OF THE RECEPTORS IN THE SKIN HAD BEEN CATALOGED. They showed that the sense of touch was far more than a single system, but was capable of detecting a wide range of stimuli.

Around the same time it was shown that the sensory nerves connecting to the spinal cord each collected information from an "island of sensibility," a specific region of the skin that became known as a "dermatome." The dermatomes vary in size, with those on the back being much larger than those on the fingertips, for example. The former is obviously less sensitive, with recepters widely spread. Dermatomes do not simply detect physical contact with the skin; they can also discern a range of pressures as well as hot and cold.

To achieve that, the skin uses six receptors, mostly named after their discovers. The Meissner and Pacinian corpuscles (also known as Vater-Pacini) detect vibrations and light pressure. The Merkel disks and Ruffini endings are slower acting and detect harder pressure. Krause's corpuscles detect cold, while heat and pain in general are picked up by free nerve endings under the epidermis.

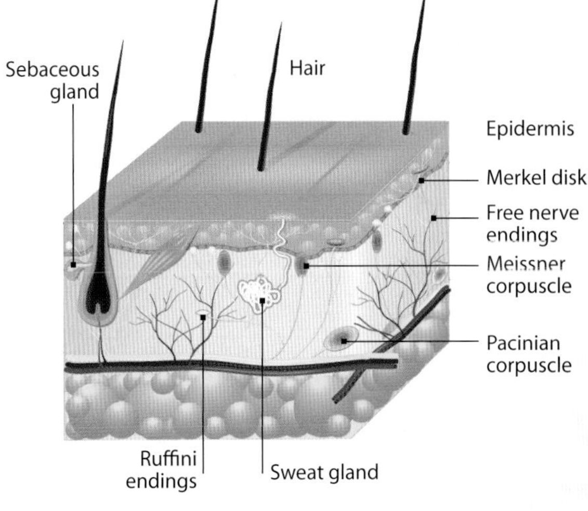

61 The Synapse

GOLGI'S DARK STAIN FINALLY PAID OFF IN THE 1890S WHEN IT WAS USED TO STUDY THE LOCATION OF NERVE CELLS IN BIRD BRAINS. What was discovered set up the neuron doctrine as we know it today: Nerve cells do not touch, but communicate across a gap, known as a synapse.

Santiago Ramón y Cajal put an end to the concept of the nerve net and ushered in our modern understanding of neurons.

For ten years, the Spanish scientist Santiago Ramón y Cajal peered through his microscope. His eyepiece was trained on brain specimens prepared using the latest techniques neuroscience had to offer: Dyed using the Golgi method and sectioned into diaphanous slices by a Gudden microtome. Ramón y Cajal was delighted with the results, especially of the staining, or "metallic impregnation" as he described it. "Everything is absolutely clear without any possibility of confusion. One need only observe and note these cells."

Sir Charles Sherrington is credited with the discovery of the synapse following his work on nerves and reflex actions at the Liverpool School of Tropical Medicine.

No matter how hard he searched, Ramón y Cajal could find no evidence of the hypothesized linkages, or anastomoses, between nerve cells. However, he did see tiny "thorns" on the plantlike branches of the dendrites. Golgi himself had seen these and dismissed them as artifacts added by the staining.

One-way street

Golgi and Ramón y Cajal never saw eye to eye on what the view through the microscope told them. Ramón y Cajal's work focused on the cerebella of birds where the nerve cells are especially large. After exhaustive observations Ramón y Cajal became certain that nerve signals (whatever they turned out to be) always traveled in the same direction through a nerve cell. He showed that the axons of sensory nerves always pointed toward the brain, while those of motor nerves faced away. Sensory nerves carry information into the brain while motor nerves take it back out to the body, and so Ramón y Cajal's conclusion was that nerve signals left the cell via the axon, and

"In view, therefore, of the probable importance physiologically of this mode of nexus between neuron and neuron it is convenient to have a term for it. The term introduced has been synapse."

CHARLES SHERRINGTON

entered it via one of the dendrites. This one-way system challenged the possibility that nerve signals flowed around, this way and that, through a continuous nerve network.

That meant, Ramón y Cajal contended, that the nerve cell, soon to be named the neuron, was "an autonomous canton." A canton, or district, often refers to a region that is self-governing, and political ideas played a role in the debates over these neurons. The Italian Golgi believed that nations should be connected in an ever larger federation, and thought that neurons did the same. Ramón y Cajal, on the other hand, thought that cultures should remain separate, but were still able to work together.

Find the gap

Despite the disagreement, Ramón y Cajal and Golgi won the Nobel Prize for their discoveries. They shared the prize with a third winner, an Englishman called Charles Sherrington. Sherrington assumed that if nerve cells did not touch, they must communicate chemically across a tiny gap. He consulted colleagues on what it should be called and agreed upon "synapse," meaning to "clasp together." Sherrington wrote of this in 1897: "If the conductive element of the neuron be fluid, and if at the nexus between neuron and neuron there does not exist actual confluence of the conductive part of one cell with the conductive part of the other … there must be a surface of separation. Even should a membrane visible to the microscope not appear, the mere fact of nonconfluence of the one (cell) with the other implies the existence of a surface of separation..."

He was right (although they were not imaged until the invention of powerful enough microscopes in the 1930s). However, chemical mechanism by which they worked remained a mystery that would take another 25 years to overcome.

If there is no connection between neurons, there must be chemical messages traveling across the synapse. It would take until the 1920s to figure out how that works.

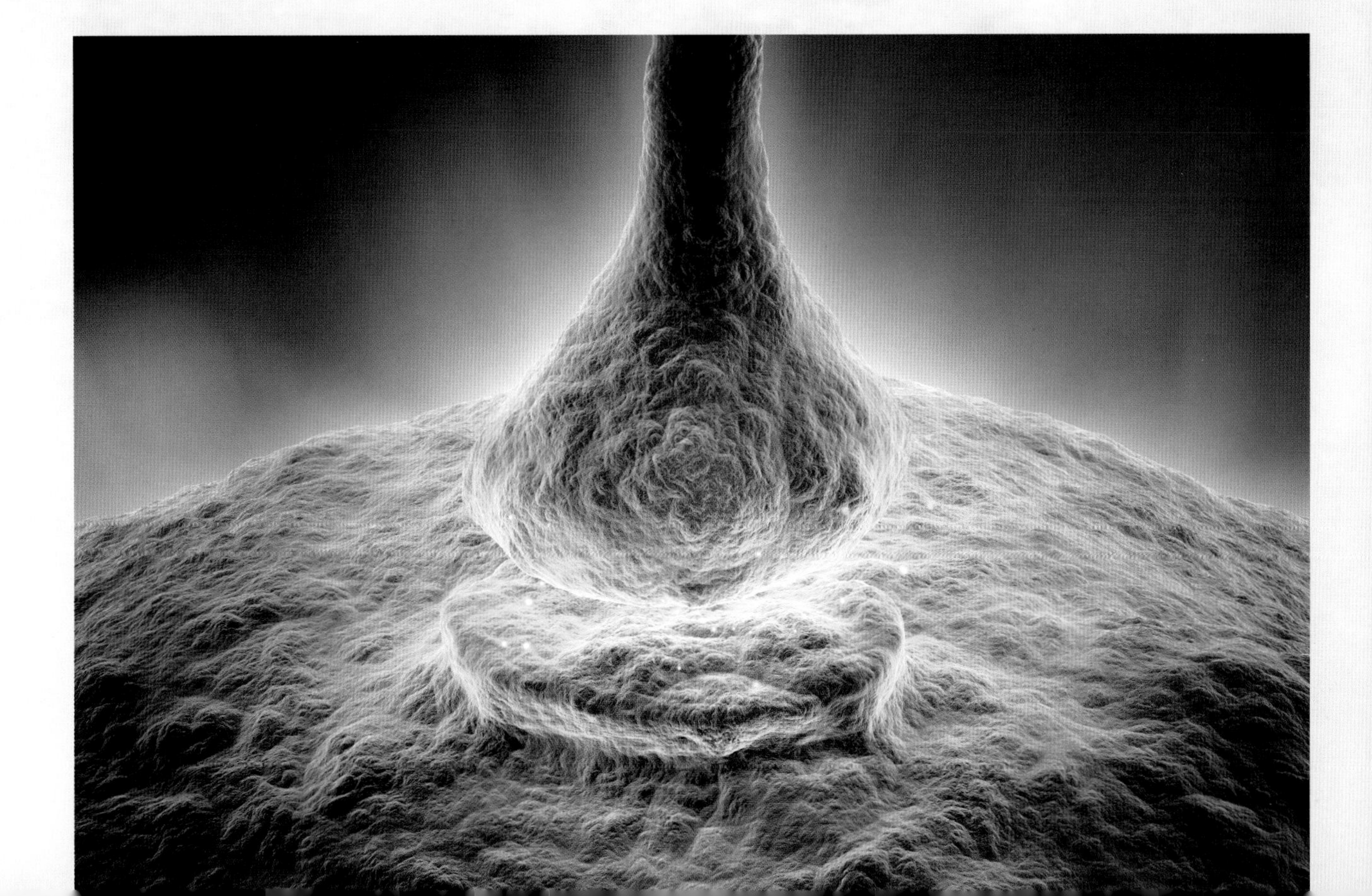

62 Autonomic Nervous System

MUCH OF NEUROSCIENCE FOCUSED ON THE VOLUNTARY NERVOUS SYSTEM, THAT BIT THAT THINKS AND TALKS and moves the body. But what of the automatic processes? Does the brain control those as well?

"The viscera is supplied by two sets of nerves ... one sets the tissue in activity, the other inhibits its action."

WALTER H. GASKELL

The body does a lot without you having to think about it. We are talking about controlling breathing and the heartbeat, and managing digestion, urination, or sweating. As usual we go back to see what the Roman doctor Galen said about it. He saw that nerves ran from the organs in the main body cavity, or viscera, and formed chains of nerve bundles that ran alongside the spinal cord. Galen thought they were "sympathetic" to the brain, carrying information about the organs. In the 1660s, Thomas Willis found that cutting the vagus nerve made the heart tremble wildly. Later researches found the visceral nerves also had effects in the face, controlling pupil size and tear ducts, and in other parts of the body. More and more, the "sympathetic" nerves seemed to control body parts rather than report back to the brain about them. In 1845, the Weber brothers were able to slow and stop the heart by electrifying the vagus nerve. Doing the same with other sympathetic nerves made the heart speed up.

A movie audience let their autonomic nervous systems let rip, as they relax in the theater but are stirred into excitement by the action.

By 1898, British physiologist John Newport Langley came up with a name for it all: The autonomic nervous system. His colleague, Walter H. Gaskell, had already found that the system was really two in one. The sympathetic nerves that emerged from the chain of nerve bundles, or ganglia, described by Galen controlled the "fight and flight" response. They prepared the body for sudden and intense activity, raising the heartbeat and increasing sensory awareness. The parasympathetic nerves, which were largely connected directly to the brain, did the opposite. They slowed breathing and the heart rate, and generally calmed the body in preparation for it to "rest and digest."

63 Bipolar Disorder

THIS CONDITION, INCREASINGLY DIAGNOSED AND FAMILIAR TO US, AFFECTS ABOUT TWO PERCENT OF US. It was differentiated from other mental disorders in 1899 by the German psychiatrist Emil Kraepelin. However, bipolar disorder and its sufferers had been around a lot longer.

Emil Kraepelin made the first study of bipolar sufferers.

It is said that the author Virginia Woolf and painter Vincent van Gogh were both bipolar sufferers. Both committed suicide. It is also said that bipolar disorder is more common among extraordinary and creative people who achieve fame in the world of entertainment and art, as evidenced by the long list of celebrity bipolar sufferers. That fact is probably hard to justify, but bipolar disorder is characterized by successive bouts of mania and its opposite depression. While depression suppresses arousal levels, mania elevates them to heights seldom experienced by nonsufferers. Perhaps mania pushes creativity into new areas as well. However, a diagnosis with bipolar should never be welcomed. Untreated, it can lead to self-harm and suicide.

LITHIUM

Sodium and potassium are natural components of the body, especially in the nervous system. No one knows why, but adding a bit of lithium, a related element, helps to control bipolar disorder.

"Though I am often in the depths of misery, there is still calmness, pure harmony, and music inside me. I see paintings or drawings in the poorest cottages, in the dirtiest corners. And my mind is driven toward these things with an irresistible momentum."

VINCENT VAN GOGH

Manic depression

The word bipolar refers to the way sufferers experience two opposite emotional states. It was introduced in the 1950s. Kraepelin's term for the disorder was "manic depressive psychosis," and he noted that sufferers generally lived normal lives, but had regular episodes of mania or depression. He monitored untreated sufferers of the disorder over long periods to build up a picture of the disease, and found that there were many variations beyond the simple up-and-down cycle. A low-level form of regular mood swings was already described as "cyclothymia," and Kraepelin found that this was the mildest form of his new disease. Most sufferers experienced more severe emotional changes and sometimes swung between the mania and depression very quickly.

Emotional response

Kraepelin was not the first to notice these kinds of symptoms. In the 1850s, two French neurologists recorded something similar. Jules Baillarger described it as folie à double forme (dual-form insanity), while Jean-Pierre Falret came up with the term folie circulaire (circular insanity).

Kraepelin suspected that the disease was inherited, and this has been born out by successive studies. People inherit a propensity to becoming bipolar, which is eventually brought on by environmental factors. One theory is that the parts of the brain associated with emotion, such as the amygdala, are more readily activated. Over time this can result in over-activity, especially during sustained periods of stress, which in turn makes the brain more susceptible to violent mood changes. Kraepelin noted that the disorder is equally prevalent in men and women, and rare in children. Another possible cause is cyclical change in the sodium channels that power neurons. Depression arises when these work slowly; mania comes from overfiring of nerve cells.

Bipolar sufferers experience long periods of normality with no symptoms, interspersed with periods of depression or mania, or a mixture of both.

64 Apraxia: Movement Disorders

"Most voluntary or most special movements, and faculties, suffer first and most. That is in an order the exact opposite of evolution. Therefore I call this the principle of Dissolution."

JOHN HUGHLINGS JACKSON

THE TERM APRAXIA RELATES TO A NUMBER OF SEEMINGLY BIZARRE DISORDERS, MOST NOTABLE WHEN A PATIENT HAS NORMAL STRENGTH BUT HAS TROUBLE MAKING deliberate movements. The study of apraxias revealed how actions were controlled by a higher level of brain functions.

Apraxias were first described by John Hughlings Jackson in the 1860s. He saw them as further evidence that the brain worked in a kind of pyramid with intellectual faculties controlling more basic functions. The subject was given greater attention by Hugo Liepmann, a German who wrote about several kinds of apraxia in 1900. He reported a patient who, when asked to comb his hair, was unable to do so, or did something else instead. However, he understood the command and knew how to use a comb. He was also able to walk and perform large-scale body movements, like sitting down and standing up—and he was also able to comb his hair when it was his own choice. Apraxias were often seen on one hand, associated with an injury to one side of the brain. Injuries to the right hemisphere led to fewer apraxias, while those on the left side, according to Liepmann, damaged the ability to organize learned movements.

65 Dementia

IN 1901, ALOIS ALZHEIMER MET A 46-YEAR-OLD PATIENT IN FRANKFURT, GERMANY. She was unable to write her name, saying: "I have lost myself". This was the first person diagnosed with Alzheimer's disease.

AUGUSTE DETER

Alois Alzheimer's medical notes for his first Alzheimer's disease patient were rediscovered in 1996. In them the doctor records how Auguste Deter appears aware that she has lost her previous abilities. Although Alzheimer moved to Munich soon after their first meeting, he made frequent visits until her death.

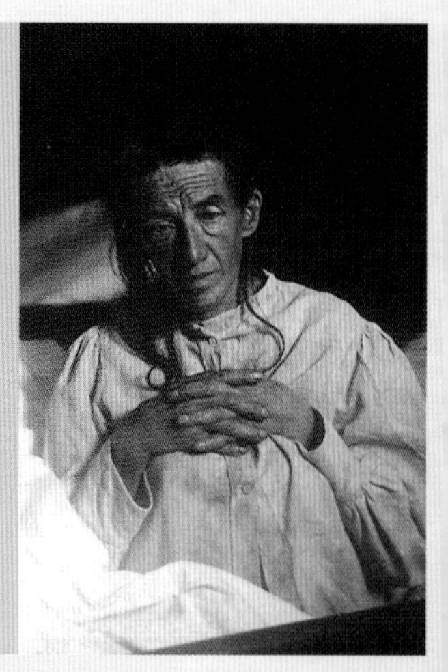

The term "dementia" refers to the gradual loss of intellect, memory, and a sense of self. It is associated with old age and takes many years to develop. As good healthcare and healthy lifestyles act to increase the average age of the world's population, dementia is likely to become much more prevalent. About a quarter of dementia sufferers have vascular dementia, where the blood supply to the brain, especially to the frontal lobes, is gradually diminished through a series of ministrokes. Nearly all of the rest suffer from Alzheimer's disease, named for the German doctor who met that

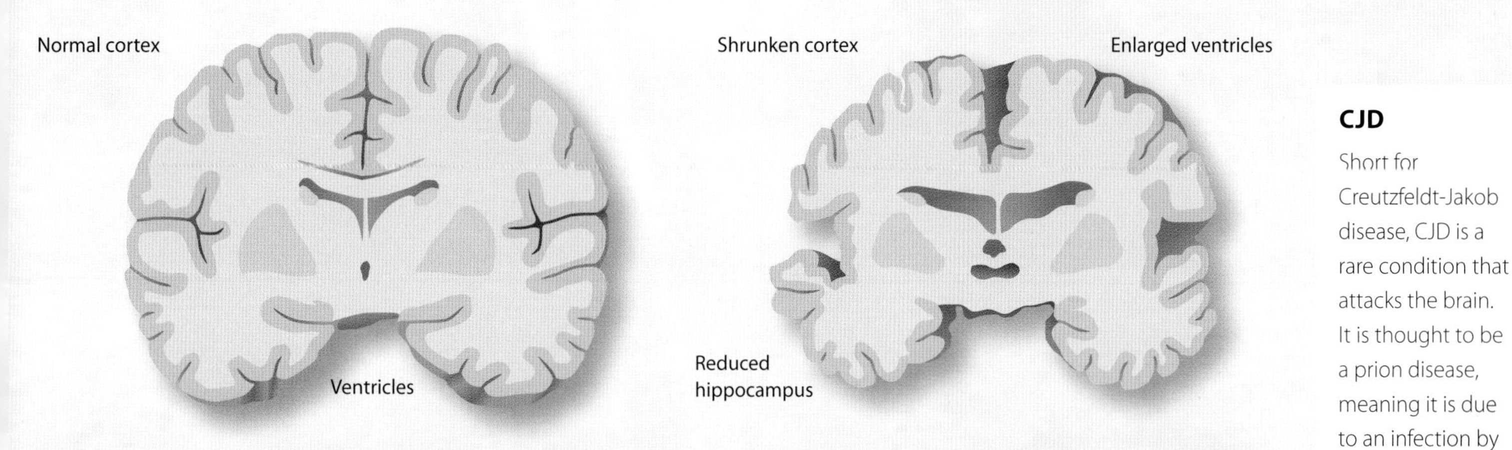

This diagram shows how Alzheimer's disease results in the cerebral cortex shrinking, the ventricles enlarging, and a marked reduction in the size of the hippocampus.

patient, named Auguste Deter, all those years ago. Alzheimer followed the progress of his patient for the next five years until she died. Auguste grew increasingly confused. She was able to identify objects but soon forgot what they were or that she had even been shown anything. Eventually, she became incontinent and bedridden, largely because she had no desire to get up. Sufferers eventually lose the ability to speak, but still have emotional responses. The muscles begin to waste due to inactivity, and eventually the weakened sufferer dies from an infection.

Search for a cure

Initially, Alzheimer's was regarded as something different to the natural senility of old age. Auguste Deter died when just 55. However, her case is now recognized as having a particularly early onset and had the same cause as the dementia that affects older people. The disease has a strong genetic component, there is no infection causing it, and there are still many theories about the biochemical causes. Without a cure, prevention techniques seem to be the best course of action. Although proof is patchy, it appears that doing regular, stringent mental activity—anything from playing a board game and performing music to reading a book or meeting friends—helps to maintain the cognitive abilities of the brain and fend off Alzheimer's disease.

CJD

Short for Creutzfeldt-Jakob disease, CJD is a rare condition that attacks the brain. It is thought to be a prion disease, meaning it is due to an infection by a protein rather than a virus or bacteria. The prion spreads through the brain, gradually destroying it. It is thought the prion comes from large mammals, such as cows and deer.

Alois Alzheimer, seated left, with his coworkers at his clinic in Munich in 1904, during the period he was building up a description of the dementia that bears his name.

66 Dyslexia

slpeling

Previously known as "word blindness," dyslexia was not really recognized as a condition until the start of the 20th century, due mainly to more children entering formal education. However, problems with writing and reading have a long history.

rdanieg

In 31 CE, the Roman historian Valerius Maximus told the story of a man from Athens who was struck on the head with a rock. He recovered fully with no loss of memory, but could no longer recognize letters. This condition was later named "alexia," and is frequently due to damage in the left hemisphere, often near Wernicke's speech area, which is involved with the meaning of words.

While alexia is the loss of reading skills, dyslexia is a problem with learning them. The first recorded case was Percy, a 14-year-old from England. Percy could only read simple words, like "on" and "the." Percy and other dyslexics (that term gradually replaced "word blind") were not unintelligent. Early investigators searched for an answer in the brain, and most settled upon a region of the parietal lobe that lies near to Wernicke's area in the temporal. Other ideas were that some left-handed readers saw words backwards. Dyslexia has a genetic component, but what is now clear is that dyslexia can arise in many ways, and therefore probably has more than one cause.

67 A Functional Map

With the debate about cerebral localization largely settled—the brain was indeed divided into units—researchers set about finding out which bits did what in ever greater detail. They did this by studying the architecture of the brain from the cells up.

The central idea behind this drive to map the brain—mostly the surface of the cerebral cortex—was that areas that did different things would have different structures from their neighbors. There was no evidence for that at the beginning of the 20th century, but it seemed like the best place to start.

Several anatomists began to make a name for themselves in this new field which had the rather unwieldy name of "cytoarchitectonics." Leading researchers included Alfred Walter Campbell from Australia, Oscar and Cécile Vogt, a French husband and wife team, and the German Korbinian Brodmann. They all followed the same broadly similar process, using stains to highlight the layers of cells that cover the

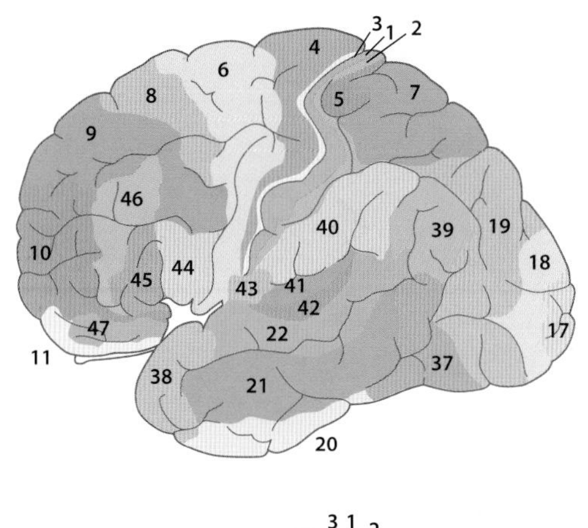

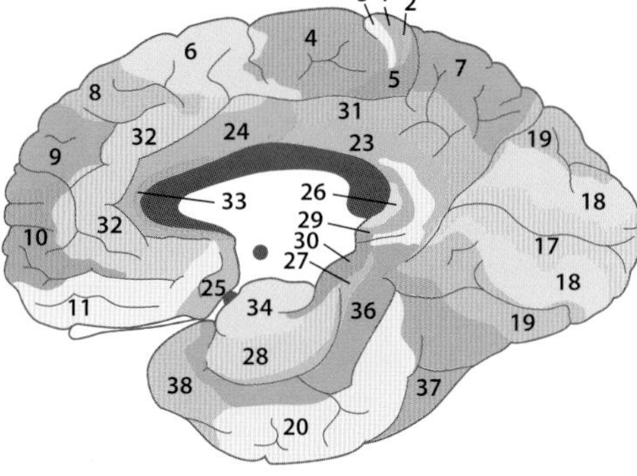

1, 2 & 3 – Touch
4 – Voluntary movements
5 – Touch processing
6 – Planning movements
7 – Hand-eye coordination
8 – Eye movements
9 – Planning and inhibiting movements
10 – Memory recall
11 & 12 – Decision making
16 – Homeostasis
17 – Vision
18 – Visual memory
19 – Motion detection and attention
20 – Face and shape detection
21 – Unclear, linked to reading
22 – part of Wernicke's area, word meaning
23 – Spatial memory
24 – Heart rate and blood pressure
25 – Appetite and sleep
26 – Episodic memory
27 – Smell perception
28–34 – Memory and navigation
35 & 36 – Memory and visual recognition
37 – Face and word recognition
38 – Unknown but linked to emotion
39 & 40 – part of Wernicke's area, word meaning
41 & 42 – Sounds
43 – Taste
44 & 45– Broca's area, speech
46 – Attention and working memory
47 – Language and syntax
52 (hidden by **41**) – Unknown

The original map produced by Korbinian Brodmann in 1909 and a list of its areas and functions.

folded surface of the brain. (If the brain's surface was laid out flat, it would cover one sheet of newspaper in an eighth of an inch of cells.) Areas of the cortex were delineated according to the arrangement of cells, the different types of cells seen and their proportions, or how they formed into layers. Every researcher had their own preferred criteria, and so the maps of the brain that began to appear throughout the first decade of the 20th century were all different. Brodmann's map had 52 areas, while the Vogts opted for more than 200 separate zones. What did these areas actually mean? Brodmann was not sure but he said: "The specific histological differentiation of the cortical areas proves irrefutably their specific functional differentiation—for it rests as we have seen on the division of labor—the large number of specially built structural regions points to a spatial separation of many functions."

Close links

In the end it worked. It was Brodmann's map, completed in 1909, that seemed to correlate most closely with distinct brain functions. The abilities of each area have been tested many times, both in the old ways of investigating the effect of damage, plus novel techniques that used electrical stimulation and the very latest in medical imaging. Brodmann's areas are now used as a road map for neurologists, forming territories on the brain. Area 17 is the primary visual cortex, Area 4 is the primary motor cortex, while Broca's area is made up of Areas 44 and 45. Higher order cognitive functions to do with comprehension and memory are also linked to certain areas. However, it is seldom as simple as all that, since each area is associated with a number of often varying functions. There was a lot more work to be done to understand the brain.

EMANUEL SWEDENBORG

This Swede had predicted brain maps 170 years earlier. He said the surface of the brain was covered in little brains, or "cerebellula," that were divided by the many folds. His ideas were remarkably close to the reality.

68 Symptoms Versus Function

Morton Prince, the neuroscientist who questioned the foundation of the neuroscientist.

JUST AS IT SEEMED THAT THE LINK BETWEEN THE BRAIN'S STRUCTURE AND FUNCTIONS was about to be solved once and for all, one neurologist pointed out that we still did not really know anything about it at all.

Since the days of Broca, the way neurologists proved the function of one part of the brain was by observing the effect caused by damage, either accidental or deliberate. Lesions in Broca's area caused a loss of speech; therefore it was concluded that Broca's area controlled speech. In 1910, echoing concerns raised by earlier researchers, American Morton Prince drew on research that showed that two-thirds of people who had lost the power of speech had a Broca's area that was perfectly healthy. That did away with the simple idea that the symptoms of a brain injury could be used to show the function of the injured region. Prince said that much of what was known about cerebral localization belonged to "the fantasies of science." He did not deny that the brain's functions were localized, but was pointing out that symptoms could not be equated with function. Each brain region relied on others to do its job, Prince said, and so brain function had to be understood as a whole as well as in parts.

69 Schizophrenia

"My patients are stranger to me than the birds in my garden."

EUGEN BLEULER

IN 1911, EUGEN BLEULER DESCRIBED A NEW FORM OF MENTAL ILLNESS. HE NAMED IT "SCHIZOPHRENIA," MEANING "SPLIT MIND." Since then the general view of this disorder has been revised. Many think that a sufferer's mind is split in two. In fact, it is split from reality.

Eugen Bleuler, an early investigator of schizophrenia.

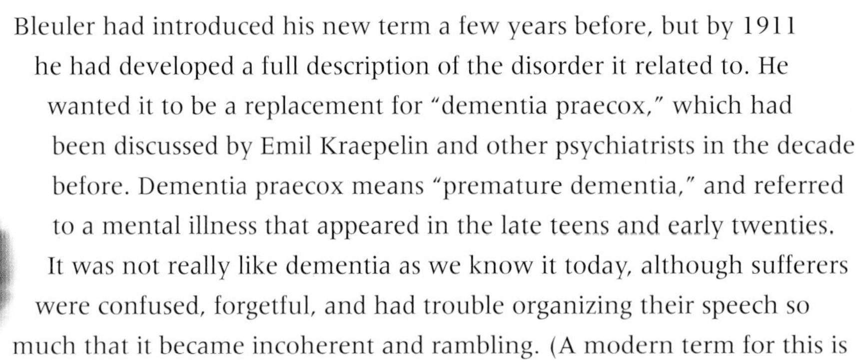

Bleuler had introduced his new term a few years before, but by 1911 he had developed a full description of the disorder it related to. He wanted it to be a replacement for "dementia praecox," which had been discussed by Emil Kraepelin and other psychiatrists in the decade before. Dementia praecox means "premature dementia," and referred to a mental illness that appeared in the late teens and early twenties. It was not really like dementia as we know it today, although sufferers were confused, forgetful, and had trouble organizing their speech so much that it became incoherent and rambling. (A modern term for this is word salad.) Bleuler recognized that these symptoms stemmed from sufferers having trouble discerning what was real from what was imaginary. Their minds had split from the reality nonsufferers shared. To a schizophrenic, hallucinations, often auditory (hearing voices) but also visual, are entirely real.

A self-portrait of a schizophrenia patient illustrates the altered state of reality they experience.

Too much, too little

Schizophrenia symptoms are characterized as positive and negative. Positive in this context means "additional" not "good." A schizophrenic experiences things that a nonsufferer in the same situation does not. They may feel, hear, and even taste things that are not there, which leads to delusions about who they are and what is really going on. A common symptom is paranoia, where a sufferer is under the impression that a hidden force is monitoring them. Negative symptoms are things like apathy, lack of emotion, or a lack of desire to maintain friendships. In extreme cases, a schizophrenic can become catatonic, either remaining still or repeating the same gesture over and over.

Schizophrenia runs in families and normally manifests itself in early adulthood. Underactivity has been reported in parts of the brain that control memory, attention, and comprehension. Sufferers also appear to be more sensitive to some of the chemicals that control the brain's nerve network. Whether this is the cause or not, most treatments attempt to correct this sensitivity.

DISSOCIATIVE IDENTITY DISORDER

In the popular imagination, schizophrenia gives someone a split personality. This is a misconception, but dissociative identity disorder does just that. Or at least some people think it does. There is no agreement on how to ascertain if a sufferer has dissociated from one identity and taken on another. The history of their other personalities is apparently lost to the sufferer, who has no idea what they did while in those states. Debate continues as to whether the disorder truly exists.

One body but multiple personalities.

70 Epilepsy

THE TERM "EPILEPSY" COMES FROM THE ANCIENT GREEK, MEANING "TO SEIZE UPON." It is the most common neurological disorder, affecting one in every hundred people, and has been recorded for millennia. It took until 1912 before medical science could do anything about it.

Many sufferers are born with epilepsy. Prince John, the uncle of the current Queen of England, died from the disease at the age of 13.

The oldest record of epileptic seizures comes from 4,000 years ago in ancient Mesopotamia. As was common back then for all neurological disorders, the problem was put down to the undue influence of the Moon god and the treatment prescribed was a bout of exorcism.

For the next 38 centuries, medical science was unable to do much better than that. In the mid-1800s, bromide was introduced as a mild sedative, which seemed to reduce the severity of siezures. The first effective treatment—phenobarbital—was discovered by chance in the early 1900s. The drug was approved in 1912 and has been used alongside more modern medicines ever since.

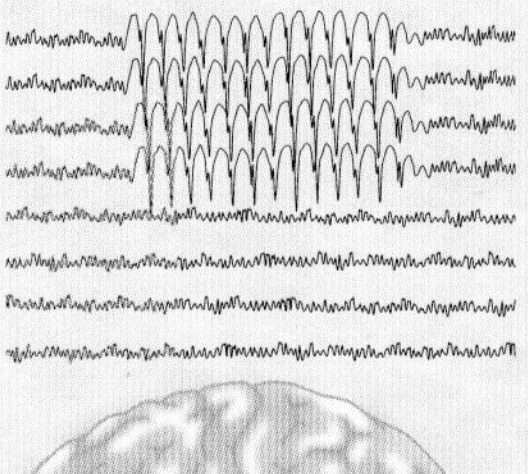

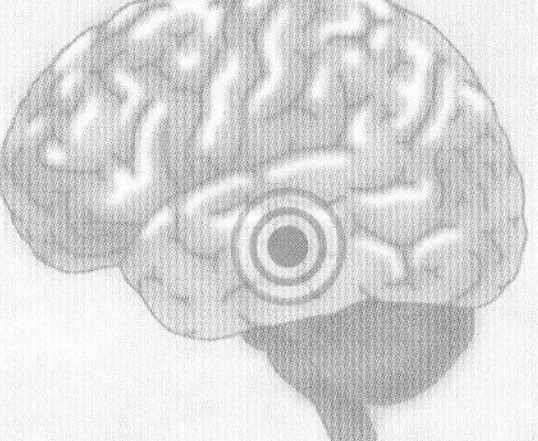

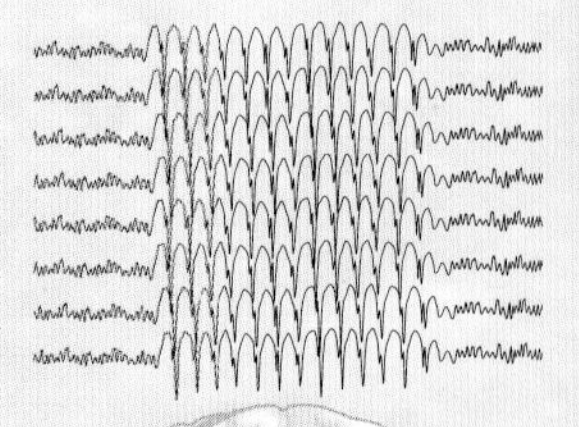

Most epileptic seizures start with activity in just one brain hemisphere, which then spreads to the other.

Status epilepticus

Until the second half of the 19th century, epilepsy sufferers were treated in the same way as the mentally ill, and were frequently shut away. Until a treatment was found, epilepsy was severely limiting to its sufferers. The Romans called it the "disease of the assembly hall," referring to the way sufferers were completely

Passersby tend to someone suffering a seizure in 1880s, London. Little could be done to protect sufferers until the advances of the 20th century. There is no cure but most sufferers can now control the problem with drugs.

symptom-free and living a normal life one minute, and then left helpless by a sudden seizure. There was no warning or pattern.

Epilepsy is not a single disease, more of a symptom of abnormal brain function that is still not fully understood. All epileptic seizures involve loss of consciousness, which poses the greatest danger to the sufferer. Most seizures are tonic, which means the muscles tense and may begin to make large, rhythmic movements of the body. Other seizures are the opposite, where the muscles loses all rigidity, and the sufferer simply collapses to the ground. Phenobarbital is most effective against the former type of attack

PHENOBARBITAL

One of a class of drugs known as barbiturates, phenobarbital was first made in 1904. Its sedative quailties were well known but its ability to prevent seizures was discovered by chance. The drug is still used for epilepsy, although medicines with fewer side effects are now more common. In the 1940s, the Nazis used overdoses of phenobarbital to murder mentally disabled infants.

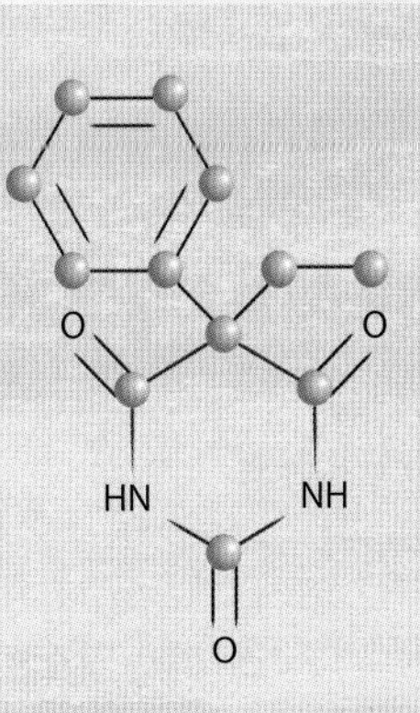

It is generally accepted that attacks that last for five minutes qualifies them as "status epilepticus." Neurological research shows that brain cells normally fire in a barely unsynchronized way. When they start to fire all at the same time, becoming more synchronized, epilipetic seizures are the result. Strokes, head trauma, and infections can lead to these kinds of seizure. Often they go away, but if they recur for no apparent reason then they are classified as epileptic seizures.

71 Nerve Center: The Striatum

THIS SMALL, STRIPED STRUCTURE IS IN THE HEART OF THE BRAIN. Early research linked it with movement control, but today it is seen as more than that—something like the brain's junction box.

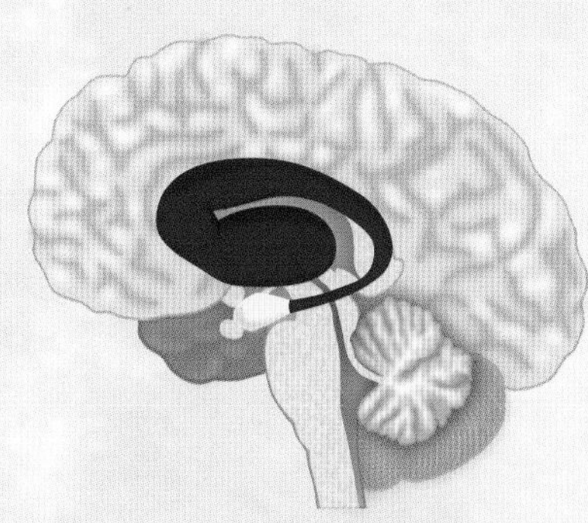

The striatum located underneath the cerebral hemispheres.

The "striatum" is named for its striated or webbed appearance. It was first identified by Andreas Versalius underneath the cerebral cortex. Thomas Willis noted two things: People who were paralyzed had very small striata, while babies did not have stripes in theirs. This led to the idea that the striatum was a control center for learned movements. Willis thought this was were the spirits of senses mingled with the ones that controlled muscles. However, when the motor cortex was discovered elsewhere in the brain in the 1870s, researchers began to study the striatum in new ways.

In 1914, S.A. Kinnier Wilson cut into the striata of monkeys and used electrical stimulation to investigate their movements. He found that the animals were still able to move but became jerky and rigid. The conclusion was that the striatum smoothed out signals to the muscles, making motions steady and controlled. This process is known as "inhibition," an important factor in coordinating the brain with the body.

72 IQ

Alfred Binet, the founding father of the IQ test.

During the early years of neuroscience the assumption was that the bigger someone's brain, the more intelligent they were. There was a pig-headed assurance that the neuroscientists themselves, and other university academics like them, were the peak of intellect—hadn't they proved that?

The idea that intellect is hereditary has a long and troubled history. Of course there is a genetic component to intelligence, but whether that alone is enough to elevate one person's intellectual clout above another's remains unanswered. Education and upbringing must also play a crucial role. But how do you go about measuring those kinds of factors? The first person to give it a go was Francis Galton, a 19th-century statistician and racist, who was an inspiration for the 20th-century's fascists.

Galton, a cousin of Charles Darwin, was a advocate of "eugenics," a process where the human race breed itself to eradicate unwanted traits. The methods would be the same as in animal husbandry—the people with desirable qualities would be encouraged to breed; those without any would be removed from the gene pool.

Measuring intelligence

Intelligence was obviously one of those desirable qualities, and Galton began an extensive research program to correlate a person's skull anatomy (broadly its size and shape) with his or her achievements. He was hoping to show that great men (and by a different standard, women, too) had bigger heads and bigger brains, capable of thinking at lightning speed. It turned out Galton's idea was a bit stupid.

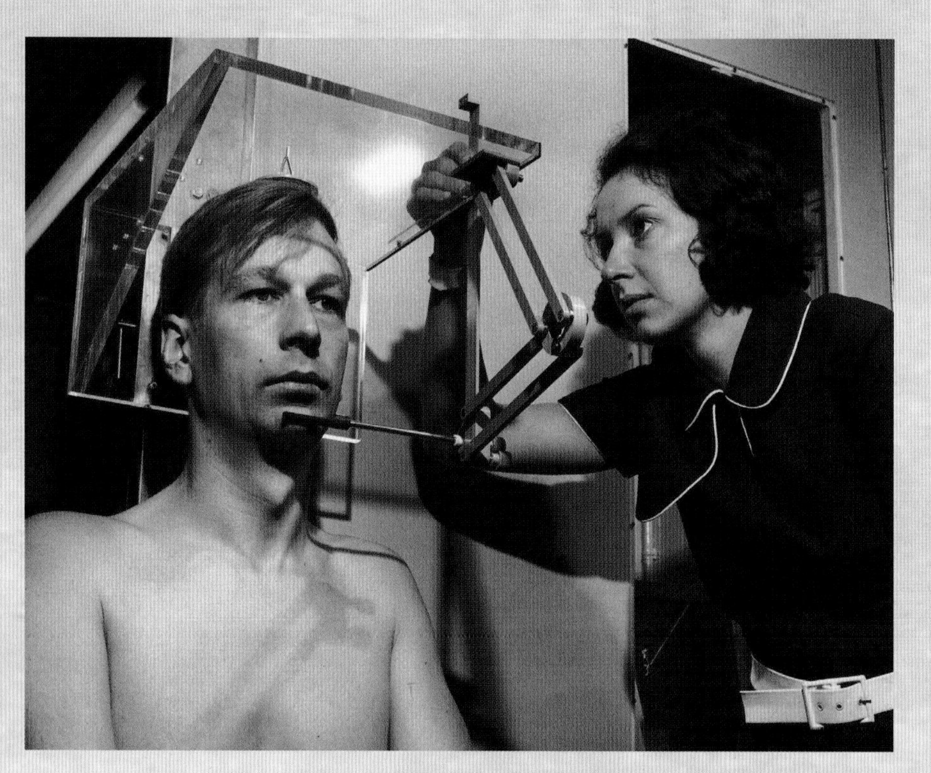

IQ has replaced measuring cranial capacity as a means of determining intelligence.

The logic of Galton and others was that intellect was shown through achievement. Around the same time, in Paris's Salpêtrière hospital, the world's leading neuroscience center, another brain researcher was feeling humiliated by a certain lack of achievement.

His name was Alfred Binet. He was a psychologist who was working for free at the hospital. (He was very wealthy and had no need of a salary.) However, most of his experimental programs were hopelessly flawed. He was out of his depth, and left in 1891. Binet then began a project measuring the skulls of clever students at the Sorbonne, Paris's great university. For ten years he was confounded by the results: The brightest students did not have bigger heads than anyone else. "The idea of measuring intelligence by measuring heads seemed ridiculous," he reported. Instead, Binet devised a test (aimed at children at first) that would show which were likely to do well in class and which would not. His test was ready in 1905 and presented practical everyday problems, such as calculating change or identifying shapes. Reading and writing skills were deliberately left out.

Standford–Binet Scale

Until his death in 1911, Binet developed his tests for teenagers and adults along with Théodore Simon. The Binet–Simon tests were used primarily in schools. Successive questions were designed to get harder, with each one deemed to be solvable by half of all people of a certain age. Therefore, the point in the tests where people began to get answers wrong indicated their "mental age," which could be actually higher or lower than their physical age.

The French tests were translated into different languages, and in 1916 a version was developed for use in North America by Charles Spearman on behalf of Stanford University. Spearman reorganized the scoring system to give a person of average intelligence (when their mental age matched their physical one) a score of 100. This system, known as the Stanford–Binet Scale, dominates the world of IQ tests today. The term "IQ," which means "intelligence quotient" arose in the 1920s and remains the shorthand for these tests today.

MENSA

People who score in the top two percent of IQs are invited to join Mensa, an international club for clever people. There are 121,000 members worldwide so it does not take a very high IQ to figure out that most people eligible haven't joined. Younger members of Mensa are cleverer than older ones. That is because we are all getting cleverer, it seems. IQ tests have to be frequently updated to maintain an average score of 100.

TAKE THE TEST

The IQ test is designed so that results fall into a normal distribution, or bell curve. Most people will get a score of 100. Higher scores suggest higher intelligence, lower scores the opposite. Only a very few people achieve scores at the far ends of scale. Just as with Binet's early versions, questions rely on nonverbal reasoning, as shown in the question below:

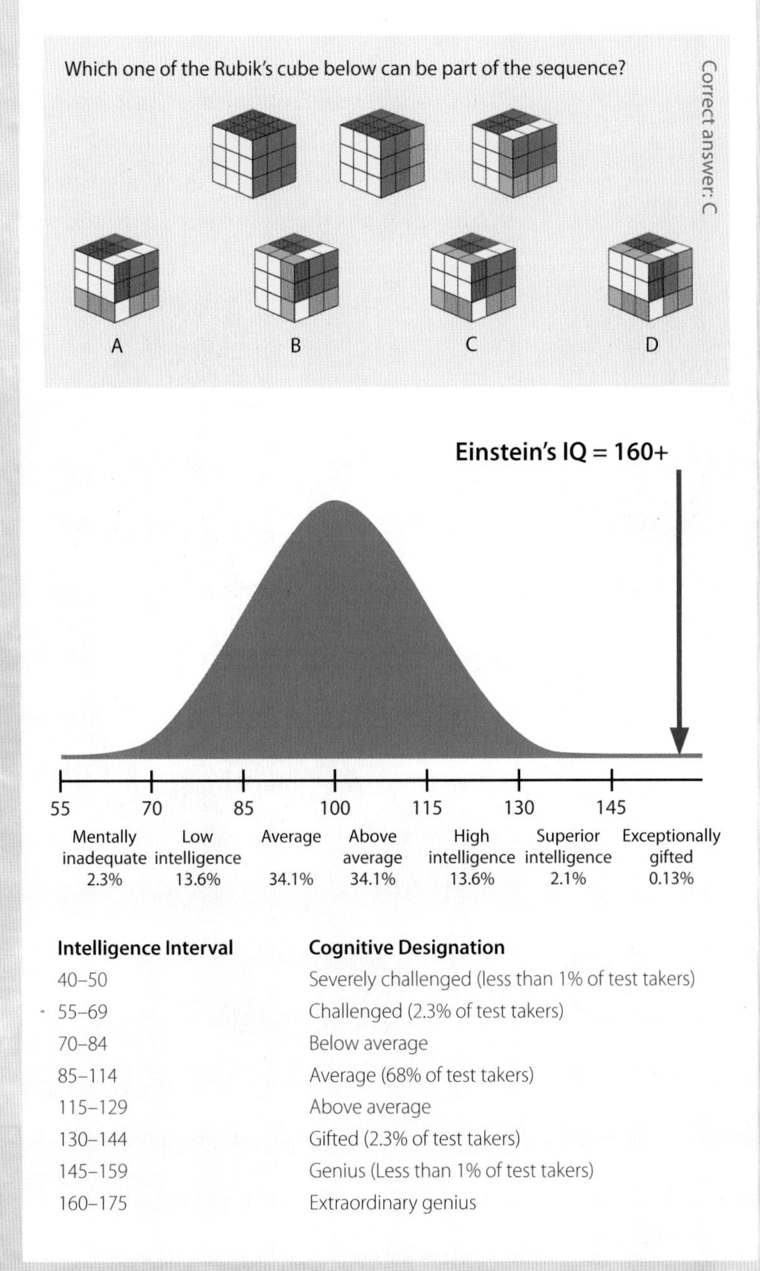

Intelligence Interval	Cognitive Designation
40–50	Severely challenged (less than 1% of test takers)
55–69	Challenged (2.3% of test takers)
70–84	Below average
85–114	Average (68% of test takers)
115–129	Above average
130–144	Gifted (2.3% of test takers)
145–159	Genius (Less than 1% of test takers)
160–175	Extraordinary genius

73 The Cerebellum

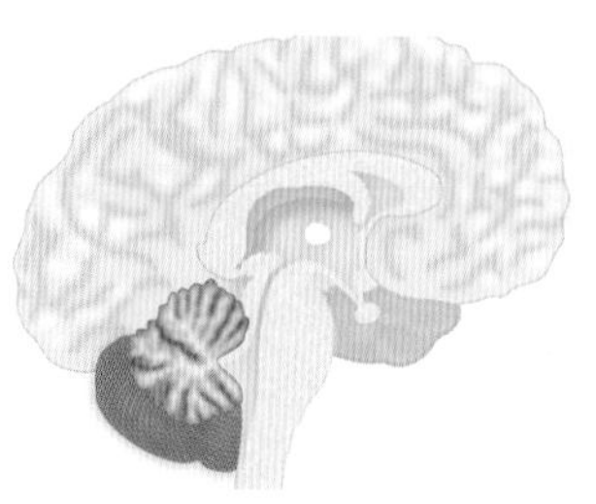

The cerebellum is underneath the occipital lobe and is separated from the hindbrain by the fourth ventricle.

LOCATED AT THE BACK OF THE BRAIN, THE CEREBELLUM WAS ONE OF THE FIRST TO BE IDENTIFIED AS A DISTINCT REGION. From the outset, it was associated with movement, and wartime brain injuries gave an insight into its role.

The word cerebellum means "little brain," and at first glance there is a lot of similarity between it and the larger cerebral lobes: It is highly folded and split into two halves—known today as "lobules." Aristotle first identified the cerebellum 2,400 years ago. Galen thought that its proximity to the fourth ventricle meant it was some kind of valve for animal spirits flowing from the brain to the body, where they create movements. For centuries, the general view was that the cerebellum controlled automatic, involuntary movements, such as breathing.

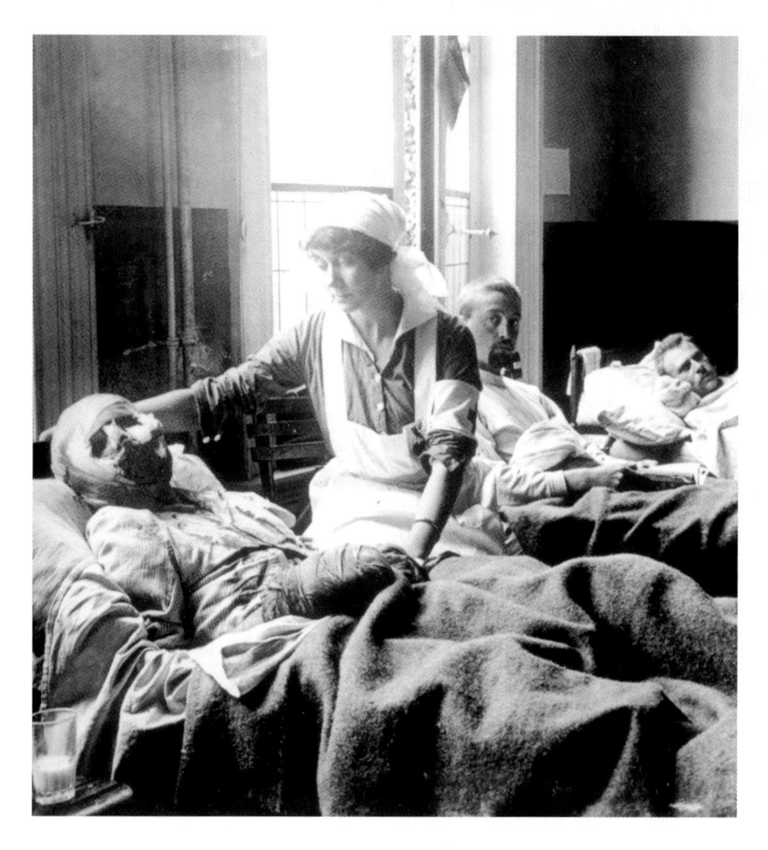

Gunshot wounds to the back of the head during the battles of World War I were used to investigate what the cerebellum did.

Alternative views

At first the cerebellum was grouped together with the rest of the hindbrain as a single unit, and it was believed that any injury to this region would stop vital motions, such as breathing. However, as anatomists began to differentiate the cerebellum as a distinct brain part, it became apparent that it was possible to survive injuries to the cerebellum and stay breathing perfectly well.

Frenchman Marie-Jean-Pierre Flourens, working in the early 19th century, discovered that losing the outer part of the cerebellum (from a pigeon) led to jerky movements; removing the middle produced constant twitching and large uncontrolled movements. If only one lobule was removed, the effect was only seen on the opposite side of the body. Removing the cerebellum entirely led to paralysis.

Battle injuries

The carnage of World War I provided ample human subjects for research into the cerebellum. A leading researcher was the Irishman Gordon Holmes. He found that injuries did not diminish the senses, and reflex movements continued more or less the same. However, he showed that the cerebellum was involved in coordinating the tension, or tone, of muscles. This allows the body to move smoothly and with the required strength. Without the cerebellum, motions are haphazard and weak.

74 The Gestalt Movement

Reducing the brain into a series of independent but connected units devoted to certain tasks was not able to answer all the questions about how the brain worked. Perhaps there was more to it than that?

As more and more regions of the brain were linked to specific functions a view formed that the organ was the sum of its parts. Some parts received input from the body and the world beyond. These were passed to other parts which processed them, moving them on to control areas, which connected to motor regions that produced a response of some kind. However, some researchers in the field found this too simplistic, especially when it came to the higher executive functions that tackled perception and cognition.

This school of thought was begun by an Austrian named Christian von Ehrenfels, who insisted that single perceptions were not enough to express the full form or "Gestalt" of the world. For example, a triangle was not perceived by connecting its three lines, and music or language were not perceived by adding together sounds in order. Instead, perceptions and other high brain functions arose from the mutual interactions of the brain's parts. This view resulted in the now well-worn adage that the whole is different (sometimes misremembered as "greater") from the sum of its parts. Ehrenfels applied this viewpoint to everything including the German nation and culture, which would have a wholly unpleasant influence later in the 20th century. However, by the 1920s, Gestalt thinking was being used to understand brain damage as an effect on the whole brain rather than the loss of one function.

"The whole is other than the sum of the parts."

KURT KOFFKA

PERCEPTION AS A WHOLE

These figures are famous examples of Gestalt thinking. Take a look, what do you see? At one level you are aware of several black shapes, all isolated from one another. However, before you noticed those your brain probably perceived something else by putting each group of shapes together into an entirely different whole.

Three pictures: A sea monster, a triangle, and a spiked sphere.

75 Neurotransmitters

Otto Loewi's discovery of neurotransmitters won him, along with Henry Dale, the Nobel Prize in 1936.

CHARLES SHERRINGTON'S PROPOSAL THAT NERVE CELLS CONNECTED ACROSS A GAP OR SYNAPSE WAS WIDELY ACCEPTED; however, no one knew how they did it. Was it electrical or chemical? A frog's heart beating in a salty liquid would give the answer.

The answer to the synapse question would come from the two basic functions of nerve cells. They either stimulate the body, making it work faster and harder, or inhibit it, making it calm down into a resting state. In the 1890s, it was found that the extract from the adrenal gland could stimulate the heart, making it beat faster. Could it be that the adrenaline produced by the gland was responsible? Did nerve signals use chemicals? The question was left unanswered because it was found that the heart was stimulated by electrifying one of its nerves—and another electrified nerve could slow the heart rate.

In 1914, Englishman Henry Dale was investigating the effects of ergot, the fungal poison responsible for St. Anthony's Fire and the medieval dancing manias. He isolated a chemical in it called "acetylcholine," which he found to have an inhibitory effect on nerves, opposite to adrenaline's stimulant effect. However, the question remained: Did the chemical influence the nerve to send a signal, or was the nerve signal producing the chemical as a means of transmitting across the synapse to a tissue?

In 1921, an experimental method for testing this question came to the German researcher Otto Loewi in a dream—actually in two dreams, but he could not remember the first one clearly enough. Loewi knew that a frog's heart keeps beating once removed from the body for some time if bathed in saline. He prepared such a heart and stimulated the nerve that lowers the rate of beating. Loewi then removed that heart and dropped in a second one with its nerves all removed, leaving it to bathe in the same liquid. Its rate dropped spontaneously. Loewi's experiment showed that the first heart had received acetylcholine from a nerve, and that same chemical left in the saline was acting on the second, nerveless heart. The mystery of the synapse had been solved: Signals are carried across the tiny gap by chemicals called "neurotransmitters." Acetylcholine and adrenaline were the first of dozens to be identified.

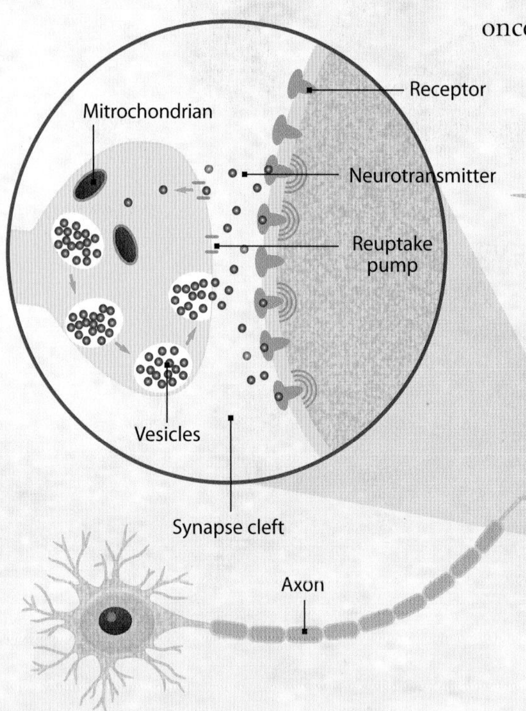

Neurotransmitters are released from one of the bulbous tips of an axon and are picked up by receptors on the neighboring dendrite.

76 Equipotentiality and Mass Action

Karl Lashley carried out exhaustive testing of rats to show that even after brain damage had wiped out their ability to remember a maze, the rodents were able to relearn it.

IN THE 1920S, TWO AMERICAN RESEARCHERS INSPIRED BY THE EXCEPTIONAL STORIES of people who recovered from brain damage formulated new theories about how the brain was able to reorganize itself to replace damaged areas and lost abilities.

While regions associated with movements, speech, and the senses had been discovered in the brain by the 1920s, no memory or intellectual centers had been located. Shepherd Franz, a researcher from Washington, D.C., wanted to know why. Most commentators thought that these functions were controlled by the frontal lobes, and pointed to evidence where damage there led to memory loss and intellectual deficits. However, during 20 years working with brain-damaged people and animals, Franz came across plenty of cases where people with such injuries had learned to walk again or regained many of their abilities to think and solve problems.

Working together

Franz teamed up with Karl Lashley, who worked in Baltimore, to investigate this using rats. Lashley would train them to complete mazes, and then send them to Franz, who cut away parts of the rodents' frontal lobes, before sending them back to Lashley. The brain-damaged rats could not remember how to run the maze, but they were able to relearn them. This led to the research team proposing two new ideas in 1929. The first was "equipotentiality," which stated that a healthy part of the brain was able to perform the role lost when another area was damaged. The second was "mass action." This put a limit on the first idea. The extent to which the brain could relearn lost skills was inversely proportional to the amount of damage. The abilities of the whole brain were impacted by the severity of an injury.

The analysis of Shepherd Franz raised the possibility that brain-damaged patients may be able to retrain the healthy parts of the brain to take control of paralyzed body parts.

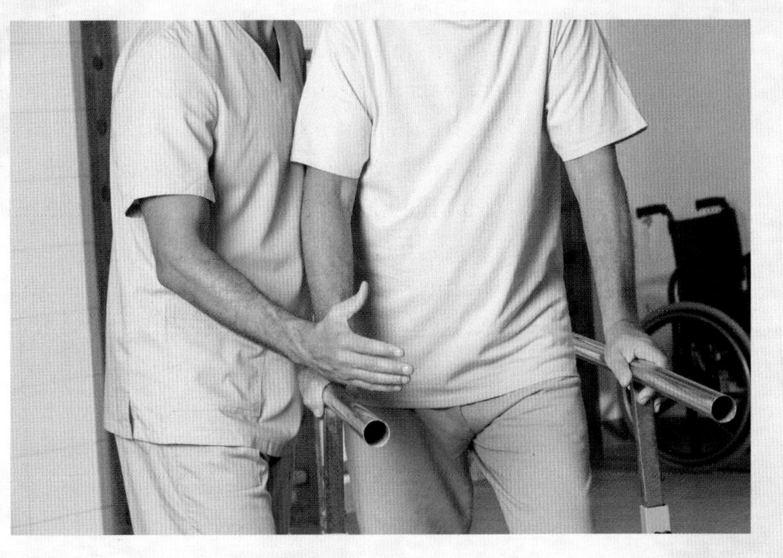

77 The Hypothalamus

STUDIES OF THE AUTONOMIC NERVOUS SYSTEM LED TO A GREATER UNDERSTANDING OF HOW THE BRAIN CONNECTS WITH the rest of the body to produce feelings of fear, joy, and rage.

The hypothalamus is, as the name suggests, beneath the thalamus that forms the base of the forebrain.

Walter Cannon, an American physiologist, began his research into the autonomic nervous system when he noticed that stomach activity was reduced when an animal became frightened or prepared to fight. He and his team found that this was caused by adrenaline that was flooding into the bloodstream. Other impacts were a rise in blood pressure and blood sugar—the body was getting ready to run from a threat or fight for its life.

Further experiments showed that these things also happened when the whole cerebral cortex was removed from animals (they used cats). This also resulted in sham rages: The cat bristled, snarled, and attacked, but was unable to focus its energies on a single threat or effect an escape as would have been the case in a real rage—and much easier with the aid of a whole brain! Cutting out the hypothalamus removed these sham rages. Electrical stimulations to the back of a hypothalamus in a healthy full brain created real rage, and thus confirmed this was the seat of the fight or flight behavior.

Primitive behaviors

The next breakthrough came from brain-damaged patients who had lost the ability to move their face voluntarily. However, they were still able to laugh and cry. It became apparent that the higher brain was inhibiting the primitive urges that were controlled in the hypothalamus. In the 1930s, Walter Cannon proposed that mental illness was caused by the hypothalamus not being inhibited correctly. However, this theory would eventually lose favor over the next 20 years as the hypothalamus was included as one part of a new, more complex emotion center known as the "limbic system."

SEXUAL DIFFERENCES

The hypothalamus is the most sexually dimorphic part of the brain: A man's hypothalamus is different to a woman's. Differences include varying sensitivities to hormones, with the male hypothalamus stimulating production of growth hormone for longer.

The hypothalamus tells the pituitary gland to secrete growth hormone. Robert Wadlow, the tallest man ever at 8 feet 11.1 inches (272 cm) had an overactive pituitary.

78 Theories of Hearing

THE FINE STRUCTURE OF THE EAR HAD BEEN WELL DOCUMENTED IN THE 1850S, BUT NO ONE KNEW HOW IT WORKED. Theories abounded as to how sound was converted into nervous signals. The big problem was the cochlea, the shell-like organ deep in the inner ear, which remained a mystery into the 1930s.

The first person to have a go at explaining how the ear processed sound was Hermann von Helmholtz. This German was a physicist by background and applied his knowledge of acoustics to mathematics to show how complex, natural sound waves could be broken down into simple units each resonating at a single frequency. In the 1850s, he proposed that the rods that Corti had found in the cochlea were all under tension, like the wires in a piano, and each was attuned to one frequency. (Helmholtz maintained that the human ear could detect 5,000 distinct tones.) The high-frequency components of a sound were detected in the base of the cochlea, while the lower frequency ones traveled up to the far end. This so-called resonance theory made sense on paper, but anatomists disputed it because they could not discern any great difference in the rods. Victor Hensen quashed the idea by showing that the endings of the auditory nerve connected to the hair cells discovered by Corti, not to the large-scale rods.

The search to understand how the ear sent sound to the brain centered on the cochlea, part of a convoluted, fluid-filled chamber in the inner ear. Also connected is the vestibular system, which works something like a spirit level and is involved in balance. If someone spins around, the fluid in the vestibular system is filled with tiny eddy currents. These take time to dissipate and result in a feeling of dizziness.

Pattern theory

In 1891, Augustus Waller introduced the idea of sounds creating a pattern in the membrane of the cochlea. This was extended by others, who described the way sounds could make an "acoustic image" as a standing wave in the cochlea, or as a ripple that moved up and down the membrane. These theories appeared to be born out by experiments. However, detractors wondered if the ear really did break down sounds only to have them recreated again in the brain.

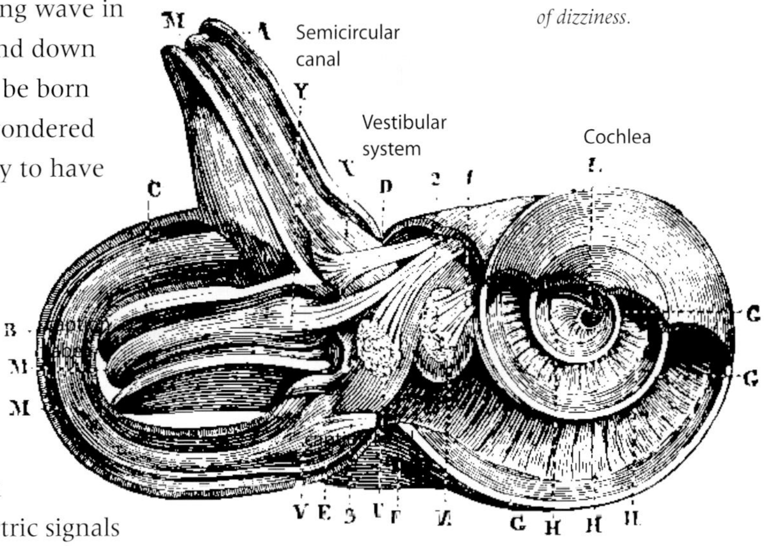

Semicircular canal

Vestibular system

Cochlea

Frequency theory

Their alternative was that every hair cell was stimulated by a sound and vibrated with the same jumble of frequencies as a natural wave. Attempts to verify this proved difficult, but in the 1930s American researchers succeeded in recording the electric signals in the auditory nerves of cats. By 1932, Leon Saul and Hallowell Davis also recorded the electrical activity of the cochlea. This showed that the frequency theory was broadly correct. In fact, the cochlea worked much like a microphone by converting the quivering motion of each hair cell into an equivalent electrical nerve signal.

79 Electroconvulsive Therapy

ELECTRIC SHOCKS HAD LONG BEEN SEEN AS A WAY OF CURING NEUROLOGICAL DISORDERS. BENJAMIN FRANKLIN GOT THE BALL ROLLING after an accidental shock knocked him over and left him with no memory of the event.

"What the Consequence would be, if such a Shock were taken thro' the Head, I know not."

BENJAMIN FRANKLIN

Franklin and others had the intuition that electrical shocks could in some way reset the brain, wiping bad memories or restarting lost abilities. From the 1780s on there were reports that this worked, but one assumes most electrotherapies had no effect at all. In 1934, the Hungarian neuropsychiatrist Ladislas J. Meduna introduced convulsive therapy for schizophrenia, where drugs were used to bring on seizures. His reasoning was that very few schizophrenics were epileptic, and therefore the two disorders were at the opposite ends of a spectrum. A seizure would bring the schizophrenia sufferer closer to the center of the spectrum and thus remove their symptoms. This was simply a guess, and an incorrect one, but in 1938 Italian psychiatrists began using electric currents to create seizures. They believed this would revitalize the nervous system. They found electroconvulsive therapy (ECT) induced amnesia, which had a positive affect on certain patients, for a short time at least. ECT is still used today to treat depression and mania, but only after other treatments have been shown to fail.

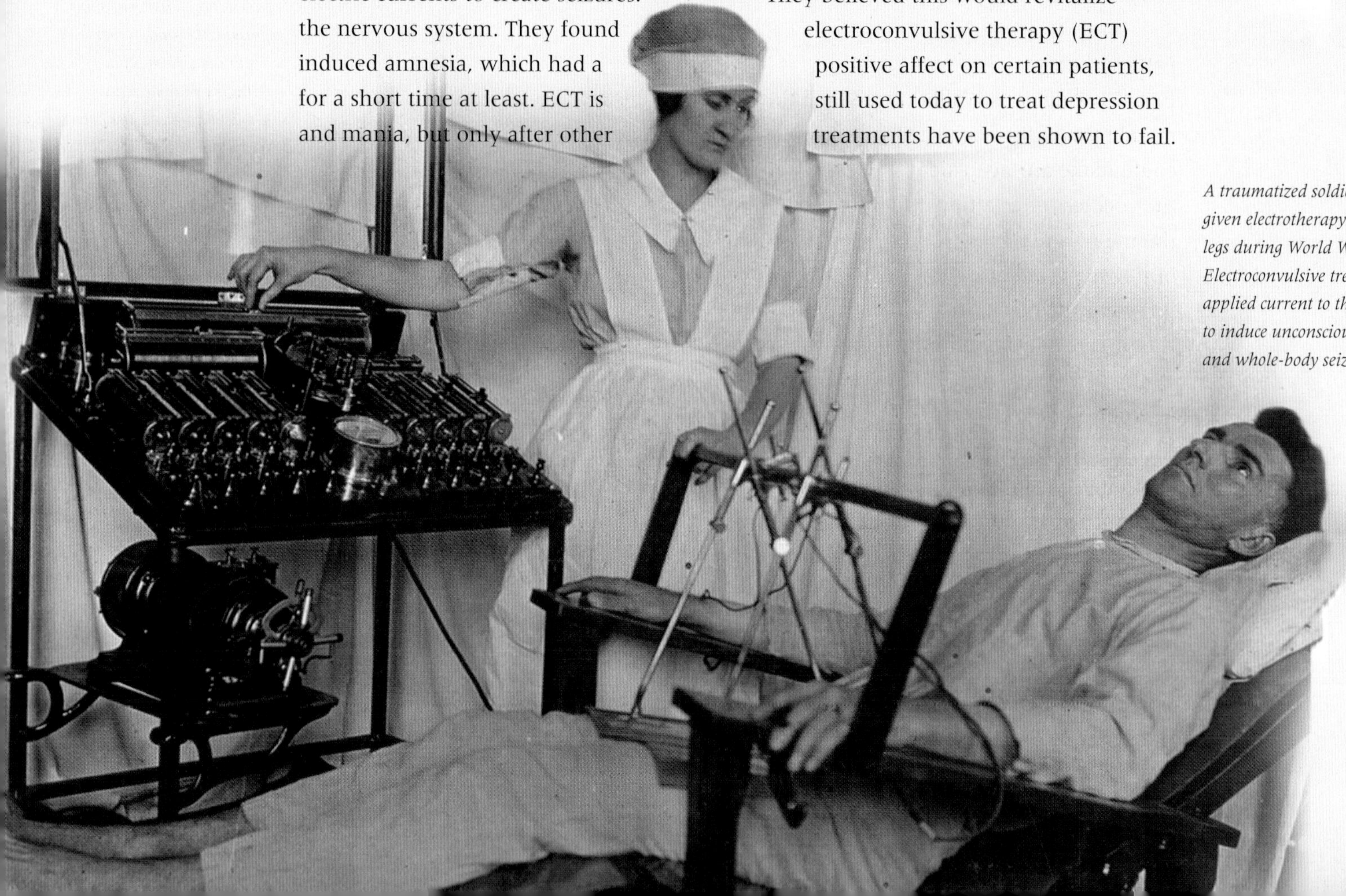

A traumatized soldier is given electrotherapy on his legs during World War I. Electroconvulsive treatment applied current to the head, to induce unconsciousness and whole-body seizures.

80 Lobotomy

While electric shock treatments were still being investigated, a Portuguese neurosurgeon was preparing for an even more radical treatment—cutting away a person's brain.

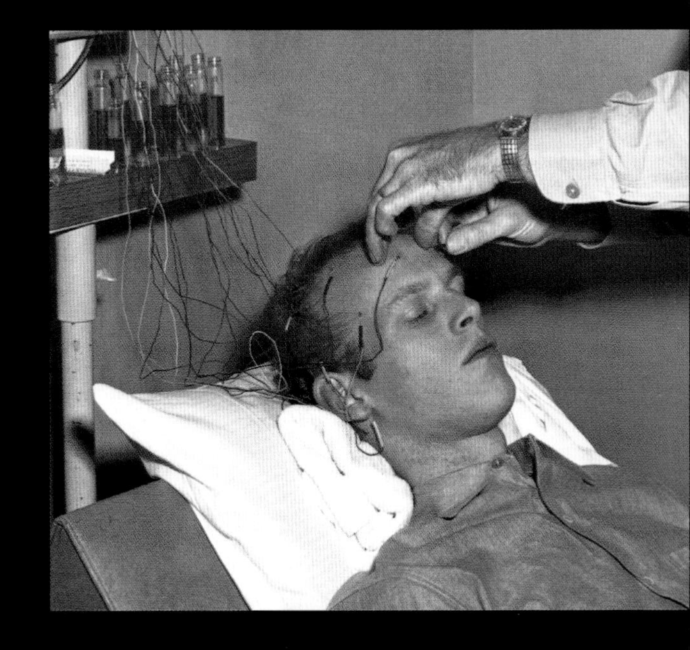

A convict in a Californian prison is prepared for a frontal lobotomy in 1961. The procedure is rarely performed today.

The surgeon in question was Egas Moniz, who was already a famous figure in neuroscience for developing a technique to image the blood vessels in the brain. In 1927, he pioneered a technique called angiography, where a liquid that is opaque to X rays was injected into the carotid artery, so it spread through the brain's blood vessels. That meant the many vessels showed up in an X-ray image, and Moniz was able to use the technique to pinpoint tumors in the brain.

Leading from the front

Buoyed by this success, Moniz was looking for his next great advance. In 1935, he had hit upon the idea that severe mental illness was caused by what he called "fixed ideas" that were lodged immovably in the frontal lobes, the seat of the intellect. His solution was hugely radical—he would simply cut out the frontal lobe. However, he needed some further evidence to back up his claim. He found it at an international meeting of neurologists that same year. There he saw how a chimpanzee that suffered near constant rages combined with low cognitive abilities was transformed into a quiet and almost emotionless creature once its frontal lobe had been removed. Despite alarm raised from several quarters, Moniz was set on performing this same procedure on his patients. In November 1935, a female patient suffering from depression and paranoia was operated on in Lisbon, with alcohol injected into the white matter in her frontal lobe to destroy all its connections.

A cross-sectional scan of a brain shows how the lobotomy procedure has destroyed most of the frontal lobe, leaving a dark empty space.

Moniz named the technique a "leucotomy," and announced that it worked well for anxious and depressed patients but had little effect on people with delusional disorders. Later surgeons developed the lobotomy, where the connections in the frontal lobe was physically severed. Within a decade lobotomy had become a quick and easy fix for severe disorders, although its benefits were overstated. By the 1950s, tranquilizers that quelled the frontal lobe were found to do the same job as a lobotomy, and the procedure gradually fell from favor.

81 Autism

A WELL-USED WORD TODAY, AUTISM IS NOW BETTER UNDERSTOOD AS PART OF A SPECTRUM OF RELATED DISORDERS. One of these is Asperger's syndrome, named for Hans Asperger, the Austrian doctor who first described autism as a distinct disorder arising in childhood.

THEORY OF MIND

Although simplistic, one way to understand autism is that sufferers struggle to understand that the content of their minds is different to what is in another person's thoughts. This is known as lacking a "theory of mind." A child can only learn to play hide and seek, for example, when the hider understands that the seeker does not share all the same information about where the players are and is seeing the world through different eyes.

The term "autism" predates the disorder we associate with it. It is derived from the Greek word for "self" and was coined by Eugen Bleuler in 1910. He used it to describe how his schizophrenia patients could withdraw into an internal world. In 1938, Hans Asperger claimed the term for another condition, one that he found in children.

Autistic behaviors

Asperger's patients were not very able to communicate through language or by nonverbal signals, and appeared to shun the outside world, adopting an "autistic" existence. They tended to spend long periods performing repetitive tasks, often ordering and stacking objects. Today, many of his young patients would be diagnosed as having "Asperger's syndrome," where the child can speak and solve problems just as well, if not better, than everyone else, although still had difficulty with interactions, and frequently becomes distressed in unfamiliar surroundings. The traditional term of autism is only really applied today to the most severe conditions, where the mental symptoms are accompanied by poor motor skills. Other sufferers are said to have autistic spectrum disorder, and a small percentage are savants, meaning they have an exceptional mental ability.

"It seems that for success in science or art a dash of autism is essential."

HANS ASPERGER

Mothers and males

Autism is most common in boys and appears in the third year of life when a child is developing its sense of self and agency. An early proposal was that "refrigerator mothers," who did not offer their children enough emotional warmth, were the cause. Today, it is understood that the disorder has many causes. One theory points to autism being more common in boys, and proposes that the autistic mind is due to an extreme "male brain," good at organizing patterns but is less able to empathize. However, critics suggest this may be due to an underdiagnosis of the disorder in girls.

82 Constitutional Psychology

William Sheldon, who had many followers in life, has been dismissed as a quack by modern researchers.

IT SOUNDS CRAZY TODAY, BUT IN THE 1940S MANY PEOPLE BELIEVED THAT THE SHAPE OF YOUR BODY INDICATED YOUR PERSONALITY. Now debunked as nonsense, the idea made sense to a lot of people at the time.

The leading figure in the field of constitutional psychology was American William Sheldon, a psychologist who happened to be the godson of William James, he of the theory of emotion. However, Sheldon owed more to the likes of Francis Galton for inspiration for his ideas. Galton believed that intellectual superiority could be discerned by the anatomy of the head. Sheldon took the belief that the physical body was linked to the mental faculties and extended it to extraordinary lengths.

Sheldon's central idea stemmed from the way a human embryo develops from three layers of cells. The digestive system grows from the endoderm; the muscles, heart, and blood supply from the mesoderm; and the skin and nerves from the ectoderm. Sheldon devised a system to score the dominance of each layer in the adult body, creating a person's "somatotype." A person dominated by endoderm—called endomorph—was physically fat, jocular, but complacent; an ectomorph was skinny, nervous, and introverted while the mesomorph was a total dish—muscular, active, and courageous.

The three principle somatotypes. Which one are you?

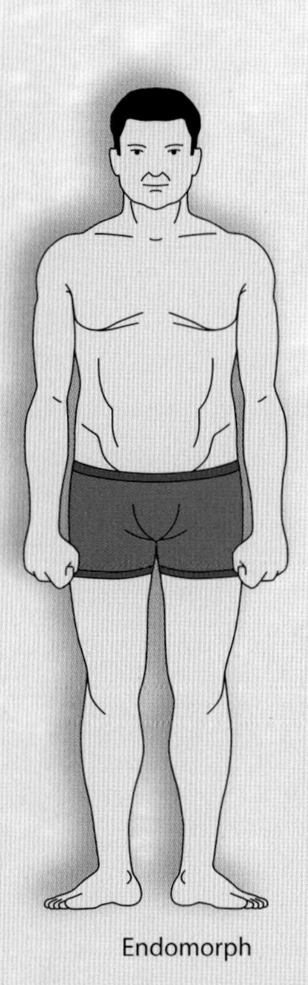

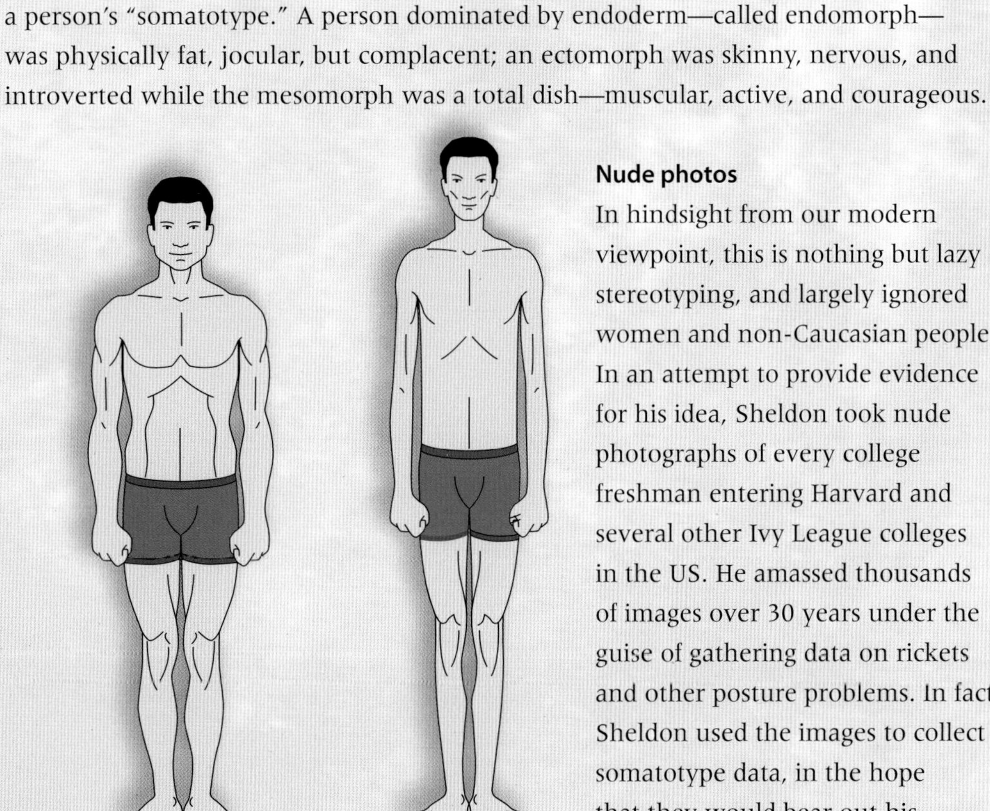

Nude photos

In hindsight from our modern viewpoint, this is nothing but lazy stereotyping, and largely ignored women and non-Caucasian people. In an attempt to provide evidence for his idea, Sheldon took nude photographs of every college freshman entering Harvard and several other Ivy League colleges in the US. He amassed thousands of images over 30 years under the guise of gathering data on rickets and other posture problems. In fact, Sheldon used the images to collect somatotype data, in the hope that they would bear out his bizarre theory.

UTOPIA

The English author Aldous Huxley was a big fan of Sheldon's idea. In his last book, *Island*, Huxley set out his ideas for a utopian society. Much of it was based on constitutional psychology with children being educated in different ways and adults being given certain jobs based on their physical appearance. This way, Huxley predicted the perfect society.

Huxley's Island *was published in 1962.*

83 The Corpus Callosum

IN THE 1940S, SURGEONS DID SOMETHING QUITE RADICAL. THEY CUT THE BRAIN IN HALF. Doing this meant severing the corpus callosum, which was regarded as the trunk line link between the brain's two hemispheres. The result was surprising: Not much seemed to happen.

The term "corpus callosum" means "tough body." It is composed of white matter estimated to contain 250 million axons that run between the brain's two hemispheres.

The surgery was intended as a last-ditch treatment for epilepsy. By disconnecting the hemispheres, seizures that developed in one hemisphere would not spread to the neighbor and take hold of the entire brain. The procedure appeared to be a success in that regard, but what of other effects? The role of the corpus callosum had been much debated over the previous century. One suggestion was that dividing the brain in this way would lead to a doubling of identity, with one-half of the mind unable to confer with the other. Evidence from clinical observations seemed to discount that possibility, but it was unclear whether that was because the hemispheres had a more intricate web of connections or because the mind just did not work that way. It was joked that the only proven function of the corpus callosum was to transmit seizures through the brain.

84 Half a Brain: Hemi-spacial Neglect

SO-CALLED "SPLIT BRAIN" PATIENTS APPEARED TO BE ENTIRELY NORMAL—IN FACT, MUCH HEALTHIER—AFTER THEIR CORPORA CALLOSA HAD BEEN CUT TO TACKLE SEVERE EPILEPSY. However, it soon became apparent that some of these people no longer viewed the world in quite the same way. In fact, they ignored half of what they saw.

One side of the brain can ignore the inputs coming from the other side.

The neurologists who assessed the split-brain patients found they did not show any reduction in intelligence or motor skills. A few had trouble speaking, but this was often temporary. However, closer inspection revealed that these people were in many ways living with two brains. Laboratory tests showed that the left hemisphere was definitely better at verbal tasks and puzzle solving, while the right one was more associated with emotional processing. Dividing the brain had revealed empirical evidence of these differences that were usually hidden by the hemispheres' constant inter-communication. A few split-brain patients, especially those who had a corpus callosum damaged by stroke, showed an especially pronounced effect.

Ignoring one side

These patients suffer from inattention, meaning they tend to ignore one side of their body and anything they see on that side. From the 1940s on, this effect was ably demonstrated by asking patients to copy a simple, asymmetrical object, like those seen on the right. A patient with damage to the right hemisphere only drew the right side of the image. The reasons for this "hemispatial neglect" are still being debated. It may be because sensory information is not reaching the executive areas, or it may indicate that parts of the brain—mostly in the parietal lobe—that are associated with focusing attention on objects are only working in one hemisphere. One remarkable example of a person with a split brain was Kim Peek (1951–2009), who was born without a corpus callosum (and inspired the *Rain Man* movie). Kim was able to read two pages of writing at the same time—and was able to commit it all to memory!

Model

Patients' version

85 The Hearing Brain

THE REGION OF THE BRAIN THAT PROCESSED SIGNALS FROM THE EARS DEFIED DISCOVERY FOR MANY YEARS. However, in 1946 its location was finally revealed—and it was in several places at once!

A cortical implant bypasses the ear and delivers electrical impulses straight to the auditory cortex.

The path to locating the auditory cortex began by tracing the nerves extending from the cochlea. They led first to the thalamus and then on to the upper part of the temporal lobe. In 1876, David Ferrier found that if he stimulated this region in monkey brains, his subjects turned their heads as if hearing a loud sound. However, the precise location proved a mystery. Studies involving cutting away the temporal lobes of monkeys failed to make subjects deaf—or at least stop responding to sounds. A new idea was to electrically stimulate the cochlea and map the corresponding activity, which showed up in the temporal lobe. In 1946, not one but two hearing centers were found there. Both processed each frequency in specific areas—although their tonal maps were arranged in reverse order to each other. Four more associated auditory areas have since been found in neighboring lobes.

86 Behaviorism

AS THE 1940S DREW TO A CLOSE, PHILOSOPHERS, PSYCHOLOGISTS, AND NEUROSCIENTISTS HAD TO COME TO AN AGREEMENT: They all had an interest in the human mind, but were they all studying the same thing?

The results from B.F. Skinner's "operant conditioning boxes" put the cat among the pigeons when it came to understanding how the mind and body worked together.

Implicit in the study of the higher, executive brain functions and especially that most human of faculties, the mind, is that a person's brain hosts private events—in short, thoughts and feelings. However, by the 1940s, this idea was under sustained attack. The problem was encapsulated in the work of Gilbert Ryle, an English philosopher, in 1949. He said that since the days of Descartes, philosophies of the mind had been laboring under a misapprehension. Descartes was a "dualist," which meant he believed that the mind and the body were separate, and this informed much of the way scientists had viewed the mind ever since. Ryle called this view "the ghost in the machine."

CATEGORY MISTAKE

One of Ryle's lasting contributions was the idea of a "category mistake." Just because it is possible to talk about the mind and the brain in the same terms, it does not mean they are part of the same category of things. That is a category mistake.

Radical behavior

The psychologist B.F. Skinner opened up another attack from a different direction. He said there was no evidence that thoughts, or cognition, was the master of the body. To prove it he developed a way of teaching pigeons and other "dumb" animals to perform complex tasks. His pigeons replicated the "clever" behaviors of apes and similar animals. No one was suggesting that the pigeons were as clever as the apes, and Skinner proposed that cognition, including that of the human mind, was not part of how the brain controlled behaviors. He believed that free will was an illusion, and that every action is the result of its consequence—we do things to get a reward or avoid a loss, just like his pigeons did. This idea is known as "radical behaviorism." The only way to counter it was to link cognition, memory, knowledge, and thought to a physical process.

Animal behaviors, like those described by primatologists such as Jane Goodall, assume that the animals have cognition, but cannot prove it.

87 The Limbic System

THE TERM "LIMBIC" IS DERIVED FROM THE LATIN FOR "EDGE," AND THE LIMBIC SYSTEM IS A COLLECTION OF BRAIN STRUCTURES THAT ARC AROUND the inside of the cerebral hemispheres. It is an informal grouping with a varying definition that has become associated with our deepest drives and emotions.

According to Paul MacLean, who led research on it in the 1940s, the limbic system includes the cingulate gyrus at the top and the hippocampus at the bottom. In between are the fornix, the mammillary body, the amygdala, and the anterior thalamic nucleus.

We have Paul Broca to thank for discovering the limbic system. He identified a great limbic lobe that surrounded the thalamus, sandwiched between it and the cerebral hemispheres deep within the brain. Broca thought it was something to do with the sense of smell, and pointed out that the same circle of structures could be seen in all mammal brains. This led to the idea that the system represented some kind of primitive, basic brain that handled the most animalistic of functions as a hard-wired emotion circuit.

Happiness unleashed by the limbic system may be the brain's own reward for successful behavior. It may also help to reinforce learning. The limbic system is also associated with the minute-to-minute short-term memory.

Three brains in one

This idea, first voiced by American James Papez in 1937, chimed with the views of John Hughlings Jackson, that animal urges lay beneath a controlling layer of humanity. The baton was passed to Paul MacLean. In 1949, he described the human brain as three brains in one: The reptilian brain was the brainstem and controlled the basic and repetitive tasks. The old mammalian brain was the limbic system, and did include the olfactory lobe as Broca had suggested. The cerebral hemispheres were the new mammalian brain and did all the clever stuff. MacLean said there were no major pathways of communication between each section and that meant they were frequently in conflict. The limbic system's job was to control the most basic behaviors, such as the drives to eat, drink, and have sex. It also turned on emotions, which prepare the body to respond to different, perhaps dangerous situations.

88 Brain Machines

IN 1949, A NEW BATTLE FRONT WAS OPENED IN THE CONQUEST OF THE BRAIN. AN ARMY OF ROBOTS WERE UNLEASHED TO SHOW JUST HOW COMPLEX nerve networks needed to be to control behaviors. The results suggested they did not need to be very complicated at all.

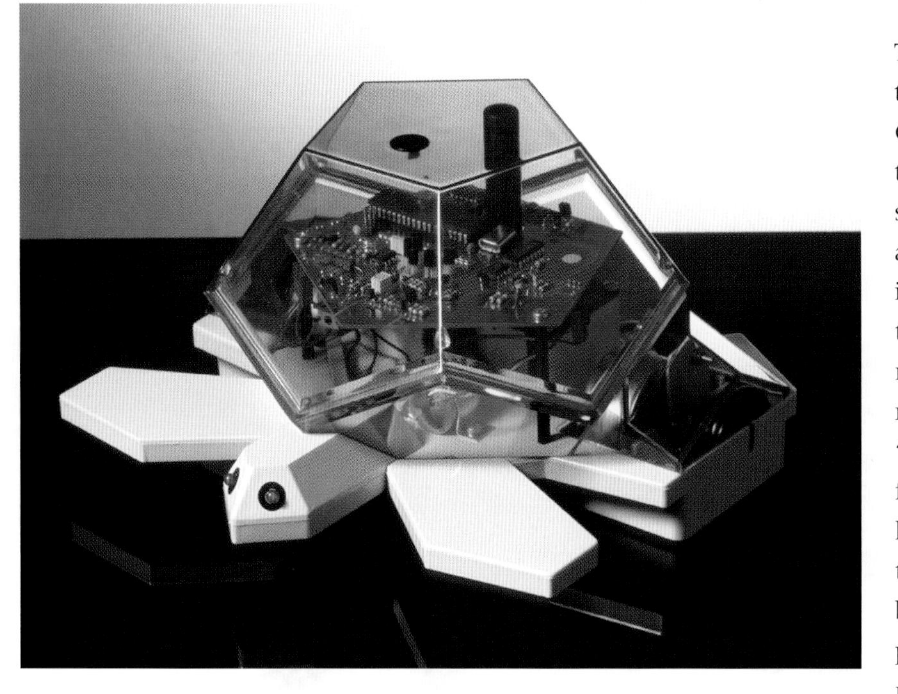

A later generation of turtle robot carries on the work begun by neurologist William Gray Walter.

The person behind using robots to study brains was William Gray Walter. His suspicion was that the rich set of behaviors seen in animals did not require a huge, complex brain. To prove it he built autonomous robots, the first of their kind. Walter's robotic helpers went by many names: He called then his "machina speculatrix," but the first two prototypes were better known as Elmer and Elsie. The type of robot being developed became known as the tortoise, partly because they "taught us" things and also because they were slow-moving and had a protecting covering. In the end, they have become "turtle" robots, and are used as toys and education tools alike.

"The mechanism of learning is of course one of the most enthralling and baffling mysteries in the field of biology."

WILLIAM GRAY WALTER

Simple behaviors

Walter gave his robots "senses." For example, they could detect light and were programmed to turn around when blocked by an obstacle. Every other aspect of their motion was random. Nevertheless, they were able to seek out the light coming from a recharging station. One robot was even fitted with its own light and found itself drawn to its own reflection—"like a clumsy Narcissus," as Walter described it.

Walter was able to teach his robots using the same techniques as used on animals in Skinner boxes. He added reflex circuits that responded to specific stimuli—pressure, light, sounds, etc.—and showed the robots modified their behaviors accordingly. These robots had a "cleverer" brain than the originals, but simply disconnecting their reflexes made them revert to more basic abilities. Walter is hailed today as a founding figure in artificial intelligence. Computers are digital—they are controlled by on/off switches. However, Walter insisted that brains understood the world in analog terms, through changing intensities of stimuli, not simply on and off responses.

89 Cognitive Behavioral Therapy

BY THE MID-1950S, A NEW FORM OF TREATMENT FOR MENTAL ILLNESS WAS TAKING SHAPE. INSTEAD OF USING DRUGS TO ALTER THE BRAIN'S FUNCTIONING, and using a mixture of discussion, counseling, and behavioral exercises to retrain the way the mind responds to certain stimuli.

Today, cognitive behavioral therapy (CBT) is the main form of "a talking cure" in that it does not involve a medical intervention, such as a course of drugs or surgery. Instead, it is a psychological therapy. CBT is frequently used alongside medical treatments, as a short-term therapy for mood-based illnesses such as depression and anxiety but also to help with more complex problems. As the name suggests, CBT is a fusion of two earlier attempts to use psychology to treat mental disorders: Behavioral therapy and cognitive therapy.

THOUGHTS create feelings

FEELINGS create behavior

BEHAVIOR reinforces thought

CBT tries to break the cycle of unhelpful thoughts and behaviors that are caused by—and then lead to—pathological emotions.

Using behavior

At its simplest, behavioral therapy attempted to apply the lessons learned inside B.F. Skinner's boxes. Therapists helped patients to identify behaviors associated with the bad feelings they wanted to eliminate. They were also advised on adopting alternative behaviors that might be less problematic. Then, the therapy would work to reward the desirable behavior and punish the bad one. This might be as simple as earning a token for good behaviors and paying a token for bad ones—although more tangible rewards were also suggested. According to radical behaviorism, the patient did not need to think about his or her problems to solve them.

Albert Ellis takes a break from some cognitive behavioral therapy, a treatment he helped develop in the 1950s.

Cognitive therapy had its roots in Freud's techniques. These practitioners searched for unhelpful ways of thinking used by patients and then sought to help them figure out a better way of looking at things. This was the initially the approach of Albert Ellis and Aaron Beck, two American psychotherapists. However, in the 1950s, they pioneered the use of both techniques at once. By the 1990s, it was eventually shown that the techniques worked better together than when used alone.

90 Action Potential

ELECTRICITY HAD BEEN SHOWN TO BE A FACTOR IN NERVE FUNCTION SINCE THE 1790S. SINCE THEN MANY ADVANCES HAD BEEN MADE WITH REGARD TO HOW THE BRAIN MADE USE OF ELECTRIC FIELDS. The neuron doctrine, which explained how nervous signals traveled from cell to cell in the brain, assumed electrical impulses were involved. But how were they made?

"The zoologist is delighted by the differences between animals, whereas the physiologist would like all animals to work in fundamentally the same way."

ALAN HODGKIN

Ever since Luigi Galvani showed how muscles and nerves were driven by some kind of animal electricity at the end of the 18th century, electricity became one of the most valued research tools in neuroscience. It lies behind the discoveries of the motor and sensory cortices, the function of the ear, and the development of the EEG. However, the exact mechanism by which a nerve cell can produce an electrical signal remained a mystery. The breakthrough would come when two English researchers began to study the nerves of the squid.

The researchers were Andrew Huxley and Alan Hodgkin, and they chose the squid as their research subject due to the animal's giant axon, a thick nerve fiber that runs the length of the body. Huxley and Hodgkin were able to study this massive nerve in great detail using a technique called "voltage clamping." Simply put, they altered the voltage of the axon and measured how that changed the way different chemicals—especially charged particles called "ions"—could move in and out of the nerve. The pair began their research in 1935, and their efforts were interrupted by World War II. As peace returned, German-born Bernard Katz made further contributions, and in 1952, the team presented their results: Most of the time the neuron did nothing, but when the time came to send a signal it used sodium, potassium, and chloride ions to create a surge of electrical potential that traveled along the axon. Once it started, the process is unstoppable and is termed the "all or nothing" response.

The electrical nerve impulse is a voltage, or "potential difference" between one side of the axon's membrane and the other. This is created by an orchestrated motion of charged particles in and out of the cell.

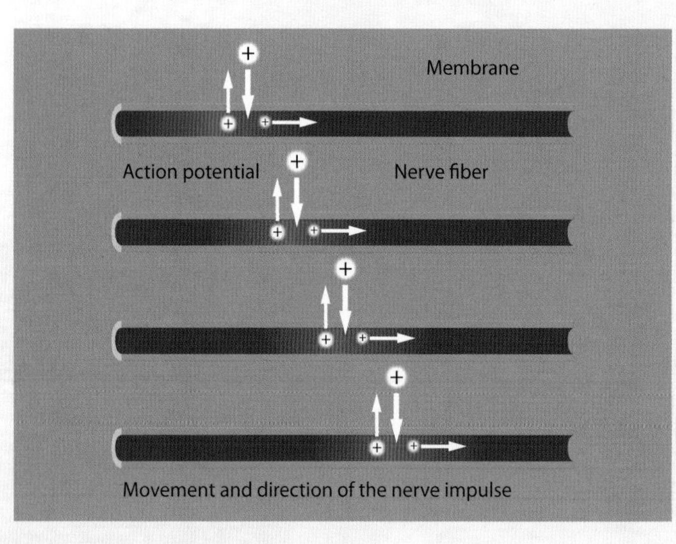

Shifting ions

When a neuron is "resting," the inside of the neuron has a negative charge while the outside is positive. This is due to certain ions being blocked from moving in or out of the cell. Positively charged potassium ions are free to move across the cell membrane in either direction, and so flood into to the cell to even out the difference in charge either side. However, negatively charged chloride ions are blocked from leaving the cell, creating a pool of negative charge inside. In addition, the axon actively pumps out positively charged sodium ions faster than the potassiums

BEGINNING AND END

The action potential is confined to the axon. It begins with a chemical change stimulated by the neuron cell body, and ends at the tip of the axon. There the signal takes on a chemical form again, as neurotransmitters move across the synapse to the next cell. The result of that may be to stimulate the neighbor to fire off its own signal—or it may inhibit it from doing so.

can get in. This takes energy, and goes some way to explain why the brain requires 20 percent of the body's fuel supply. It needs feeding even when its doing nothing at all (which it never really does, of course).

Into action

The resting potential of the axon is –70 millivolts. When the axon is asked to send a signal current, a chemical stimulus from the neuron causes channels to open in the axon's membrane close to the main part of the cell. These channels allow sodium ions to flood freely back into the cell. The potential begins to change in this part of the axon, and if it gets above a critical level (–55 mV), the effect will suddenly accelerate with more sodium channels opening. The polarity of the axon flips as the outside becomes negative and the inside positive. After this sudden switch in potential has occurred, the system then reverts to normal with sodium ions being pumped out of the cell to restore the resting potential. However, the effect of the changing polarity leaks along the axon, repeating the process over and over to create the spike of "action potential." The spike is the nerve impulse. Although made from electricity, the impulse is not moving at the speed of light like an electrical current. The velocity varies, being faster in white matter, but is between 3 and 400 feet/second.

The action potential of a motor nerve is what causes muscle fibers to contract. The contraction process is similar to the one used by neurons in that it relies on the movements of charged ions. If those ions become scarce—perhaps used up in sweat—then the muscles begin to cramp up.

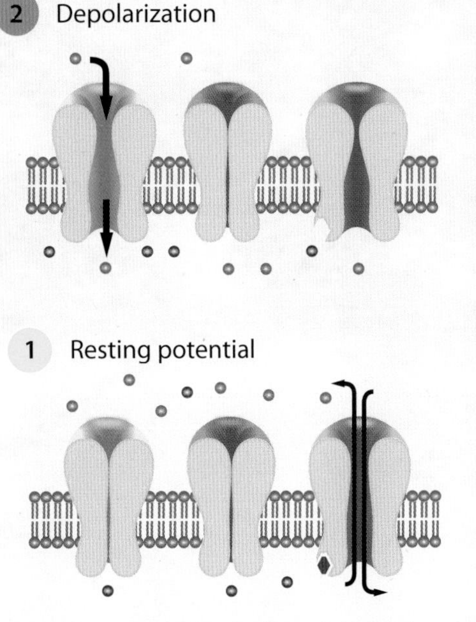

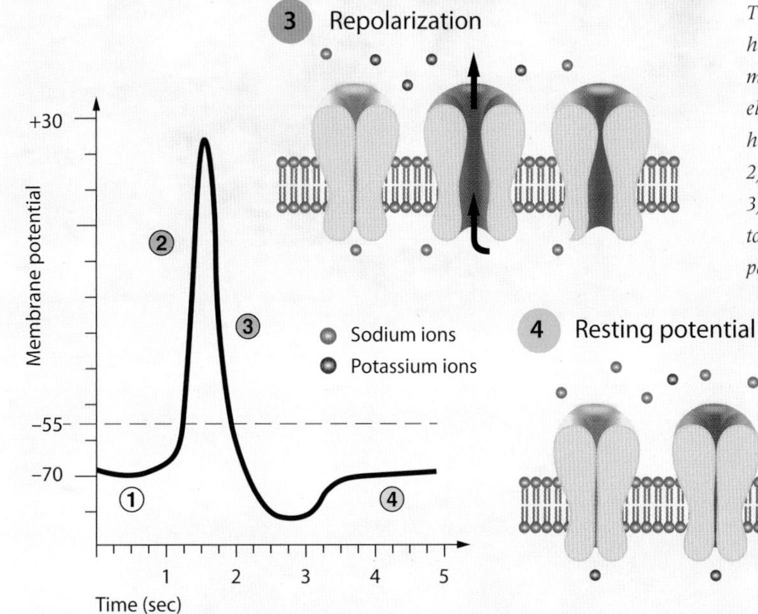

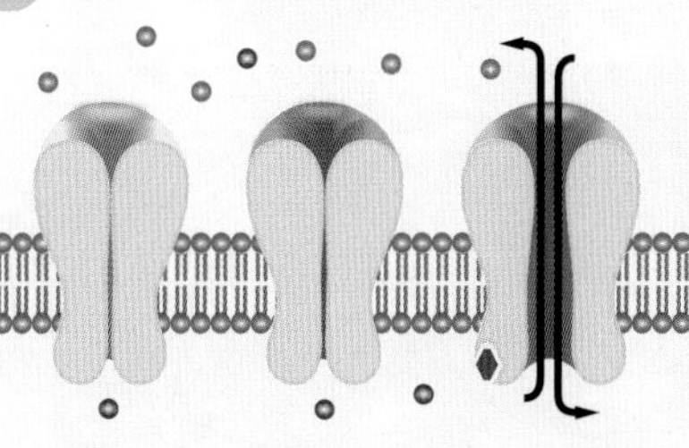

This four-step diagram shows how ions move across the axon's membrane to create a spike in electrical potential. 1) The membrane holds the concentration of ions steady. 2) Sodium ions flood into the axon. 3) Potassium ions are pumped out to repolarize the cell. 4) The resting potential returns.

91 The Sleep Cycle

FOLLOWING REPORTS THAT YOU COULD TELL IF A PERSON WAS DREAMING BY THE WAY THEIR EYES MOVED UNDER THEIR EYELIDS, A STUDENT FROM THE UNIVERSITY OF CHICAGO wanted to find out for sure: He stayed up all night watching people's eyelids. Then his supervisor got involved by recording the brain activity of sleepers. Together, they discovered the sleep cycle.

"I fellowed sleep who kissed me in the brain,

Let fall the tear of time; the sleeper's eye,

Shifting to light, turned on me like a moon."

DYLAN THOMAS

The research team was Eugene Aserinsky, a physiology student, and his boss Nathaniel Kleitman. Their interest had been piqued by the assertion of Edmund Jacobson, who, in 1938, had written a book about how to have a good night's sleep. Jacobson said that dreaming was accompanied by the movement of eyes, and he'd made photographic records. By 1953, these records had been lost, and so Aserinsky started from scratch. His resulting thesis is seen as the first evidence of REM, or rapid eye movement, sleep. In later research, Kleitman showed that during REM sleep, the brain waves recorded by an EEG were similar to the ones seen in a waking person. When sleeping test subjects were woken from REM sleep they reported being in the middle of a dream—even those people who declared they never had dreams. Kleitman and Aserinsky showed that everyone dreams; it is just that some people do not remember them.

SEEING GHOSTS

Around 40 percent of people experience sleep paralysis at some point in their lives. They wake up but are unable to move. This is because they have woken during REM sleep when the muscles are paralyzed. On rare occasions, people report seeing ghostly figures during sleep paralysis. One theory is that this is an error in the way the brain builds a model of its body. Unable to detect the paralyzed body, the brain mistakenly places it in the wrong location and creates a ghostly hallucination.

Sleep laboratories

The researchers, who by now included William Dement, developed a set of instruments for monitoring subjects in a sleep laboratory. They

recorded heart rate, body temperature, eye movements (left and right), snoring levels, the electrical activity in muscles, as well as monitoring how brain waves changed. The EEG record from a single night's sleep took up half a mile of paper!

Kleitman and co. used this equipment to figure out why some people had trouble sleeping—or slept too much. They even took it down into the darkness of Mammoth Cave in Kentucky, and under the sea in submarines to investigate the effect of sunlight on sleeping patterns.

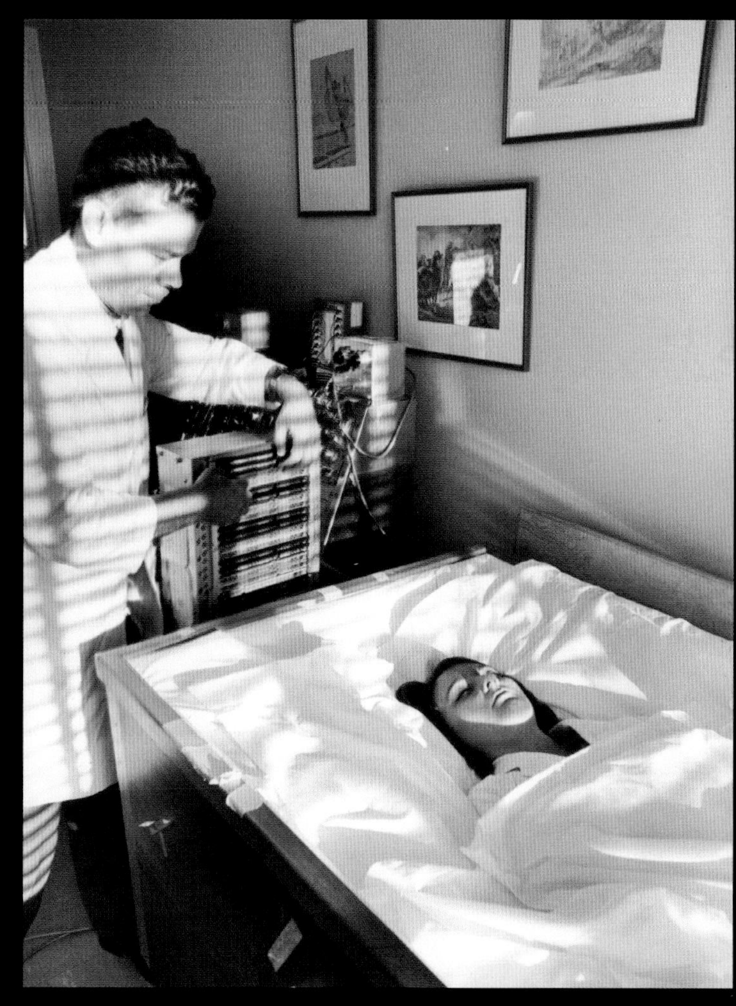

A sleep lab from the 1970s suspended subjects in a water bed to create a feeling of weightlessness, and tested how well people slept.

Sleep laboratories

The research revealed that every night's sleep followed a pattern, known as the sleep cycle, broadly made up of periods of NREM (non REM) sleep punctuated with REM sleep. It runs for 100 minutes and so there are several cycles in a normal night At the start, NREM is the dominant state, but as the night goes on, periods of REM get longer and longer. Some neurologists have suggested that there are three brain states: Wakeful consciousness, REM, and NREM. (Another idea is that hypnosis is another distinct state, although there is no real agreement there.)

The cycle is made up of distinct stages. Stage 1 lasts about ten minutes. It is when you have just dropped off, and you can be woken easily. The muscles are still able to move and may twitch, your eyes roll around and open and close occasionally. The brain waves change from the alpha waves of resting wakefulness into the theta waves of sleep. In Stage 2, theta waves dominate, and the muscles relax. It is quite hard to wake you up. This state makes up about half of the total sleep in adults.

Stage 3 (some authorities divide it into 3 and 4) is deep sleep. This is when you are getting a good rest, and barely respond to external stimuli. Most deep sleeping is done in the first sleep cycle of the night and can last about 45 minutes. At the end of Stage 3, about 90 minutes from going to bed, REM sleep takes over. The eyes start to move, but the rest of body is more or less paralyzed—to prevent it acting out dreams, perhaps. Brain waves begin to resemble those of wakefulness, and the heart and breathing rate begins to fluctuate. Ironically, this phase is when a person is hardest to wake. The dreams last about ten minutes, and then you fall back quickly to Stage 2 (via Stage 1 if the dream woke you up). As the night continues, the length of REM sleep increases and deep sleep diminishes to nothing as the body prepares to wake up.

92 The Memory Trace

BEHAVIORISTS POSED A CHALLENGE TO NEUROSCIENCE BY SAYING THERE WAS NO MENTAL COMPONENT TO LEARNING AND THAT OUR CONSCIOUS DECISIONS WERE ILLUSORY. Few believed this was really the case but to prove it science needed to uncover the physical part of memories and knowledge.

Eric Kandel won the 2000 Nobel Prize for his work on the chemical activity associated with learning and memory.

Although no one knew what it was, the physical part of a memory already had a name—an "engram." One of the first researchers to reveal something about this hypothetical entity was Karl Lashley, the American neuroscientist who, with Shepherd Franz, had revealed the plasticity of the brain. Lashley's work showed that an engram was not stored entirely in one place, and when partially damaged (by removing part of the brain), other parts of the brain were able to connect with the residue of memory and restore it. This led to the idea that a memory was stored as a distributed network of nerve cells.

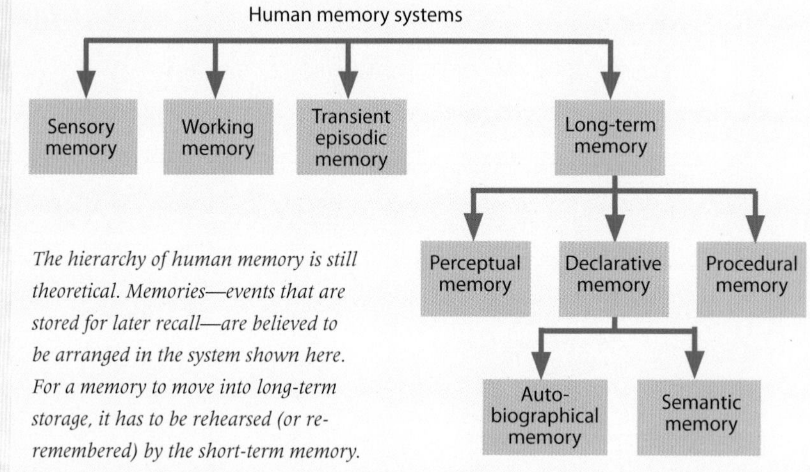

The hierarchy of human memory is still theoretical. Memories—events that are stored for later recall—are believed to be arranged in the system shown here. For a memory to move into long-term storage, it has to be rehearsed (or re-remembered) by the short-term memory.

Fire together

In 1949, Donald Hebb, an Canadian psychologist, proposed a mechanism by which "memory networks" formed: "Cells that fire together, wire together." In other words, repeated signals between a collection of nerves strengthened their connections. In the act of learning something new, the brain built a network of cells that persisted as a memory—as long as it was recalled or used enough to maintain the links between cells.

However, all this was theoretical until the work of Eric Kandel in the late 1960s. Working at a government medical laboratory in Maryland, Kandel studied the chemical activity of nerve cells in a sea slug. He uncovered the chemical changes that occurred when the network of neurons was "remembering." This work provided the first evidence of Hebb's theory of learning.

Sea slugs are just as capable of learning as a human and they use the same biochemistry to do it!

So what was the brain remembering? It is thought that the brain has several memories: The sensory memory retains information for less than a second, while the working memory holds details of the here-and-now for about a minute or so. The episodic memory is similarly short term and is a chronological record of your life. From these short-term databanks, some memories are transferred for long-term storage and are arranged as facts, events, and procedures.

93 Coma

SERIOUS BRAIN INJURIES MAY CAUSE COMA OR AN EXTENDED PERIOD OF UNCONSCIOUSNESS. In 1974, a testing regime was devised to measure how deep the coma is, and thus throw light on its possible causes.

A person is said to be comatose when they are unconscious and cannot be woken even by loud sounds and painful stimuli, like pinching and needle pricks. Their brain is not running the normal sleep cycle of deep sleep and dreaming, and they do not adjust their body posture. Coma can be caused by a brain injury, but it is more often the result of poisoning—mostly from illicit drugs—and oxygen starvation.

GLASGOW COMA SCALE

BEHAVIOR	RESPONSE	SCORE
Eye opening response	Spontaneously	4
	To speech	3
	To pain	2
	No response	1
Best verbal response	Oriented to time, place, and person	5
	Confused	4
	Inappropriate words	3
	Incomprehensible sounds	2
	No response	1
Best motor response	Obeys commands	6
	Moves to localized pain	5
	Flexion withdrawal from pain	4
	Abnormal flexion (decorticate)	3
	Abnormal extension (decerberate)	2
	No response	1
Total score	Best response	15
	Comatose client	8 or less
	Totally unresponsive	3

Keeping score

With the patient beyond the reach of clinicians, doctors turn to other evidence to figure out what is happening. The main system used is called the Glasgow Coma Scale, which was developed in the 1970s. It gives patients scores for their responses to light and sound and the ability to use their muscles. A score less than 8 means you are in a coma. Below 3 and you are clinically dead. Doctors also look out for "locked-in syndrome," where an injury to the brainstem can return a coma score, even though the patient is conscious.

POSTURES

Coma patients frequently adopt abnormal postures. These are the result of pressure in the brain following an injury. While each posture poses problems for caregivers, they also indicate where and how serious injuries are in the brain. The decorticate, or "mummy," posture suggests a problem in the cerebral hemispheres or midbrain. The decerebrate posture is more serious—it scores lower on the motor response scale—and suggests damage to the brainstem. A third posture, opisthotonus, where the whole body curls backward, is linked to the spinal cord.

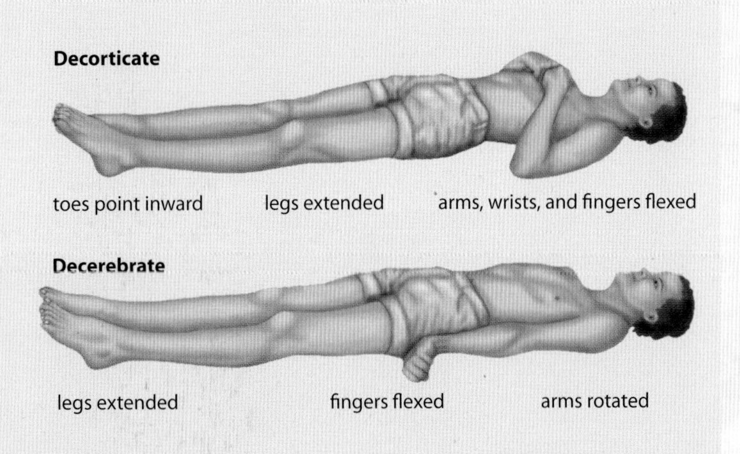

94 Positron Emission Tomography

MANY ADVANCES IN IMAGING, SUCH AS ANGIOGRAMS AND EEG, COULD SHOW BRAIN STRUCTURES, but until the 1970s no scan showed which bits of the brain were active. Then came the PET scan.

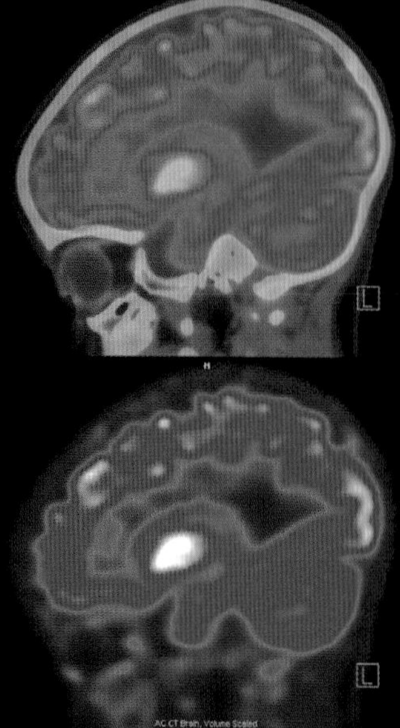

The brighter regions indicate that glucose is being used to power that part of the brain.

The PET scanner was put into service in 1976. It requires making the bloodstream mildly radioactive by injecting biologically active chemicals that carry an isotope emitting positrons. These radioactive tracer molecules are chosen for their association with a certain organ. For brain scans, glucose is often used. The marked glucose is given time to gather in the brain and then the scanning begins. The positrons collide with electrons making a flash of gamma rays, which the scanner picks up. Each scan shows where in the brain the glucose is being used. The latest scanners can model the brain in 3-D, but the real breakthrough was that PET scans revealed the locations of brain functions in living patients.

95 Identity

COGNITIVE SCIENCE IS A BROAD FIELD OF STUDY THAT MERGES NEUROSCIENCE WITH PHILOSOPHY AND COMPUTER SCIENCE. IN THE 1980S, philosophers returned to old questions about how the mind was aware of itself, but this time they asked neuroscience to help them find an answer. A good starting point was the concept of personal identity.

The years take their toll, but is it the same person inside each of these heads?

ROBOT TWIN

Hiroshi Ishiguro was born without a twin, but he has made himself one—a robot version that he calls a "geminoid." The Japanese engineer often sends his geminoid on business trips to represent him at meetings. The robot provides the body (which is constantly refined to look like Ishiguro) and video conferencing technology allows the human twin's mental component to be present at proceedings.

In the 17th century, the English philosopher Thomas Hobbes described a puzzle known as Theseus's Ship. Theseus, the Greek hero, needs to repair his ship. Old decking, ropes, and sails are torn out and new ones are put in. The ship is so old, every one of the original elements is removed and replaced. Another shipwright, meanwhile, collects all the old pieces and reassembles them into an entire ship. Both set sail. Which one is Theseus's ship? You might say the one made from the original materials, but Theseus is on board the new one, and he disagrees. Hobbes is pointing out that identity is not defined by physical substance. Winding forward almost 400 years we could ask the same question about your body, which is consistently renewing its skin, blood, and many other body parts. Is it still the same body as the one you were born with?

Mental content

In 1984, the British philosopher Derek Parfitt reran the question with regard to human identity. He imagined a machine that can transport a human body from one planet to another—but something goes wrong and two identical humans arrive. Which one is the original? Do they both have the same identity? Parfitt was drawing on John Locke's ideas about identity, and assumed (as most of us do) that the identity of both teleported astronauts comprised a collection of the same autobiographical memories. They were both the same person—until the moment they met, that is.

Parfitt's thought experiment was devised to show that identity—our sense of self—only exists in the present, and changes with every moment of our lives. However, we look upon it as a continuous stream of being. Parfit suggests this is an artifact of our survival instinct, which draws upon our memories of past events to help it tackle problems in the here-and-now and beyond. This process is what connects our history. We only know ourselves by remembering what we have achieved rather than there being a central store of primary facts that define us. However, does this fit with evidence from neuroscience?

Judgement call

Brain scans suggest that our sense of self is formed in the medial prefrontal cortex right at the front of the brain. When test subjects are asked to judge themselves (e.g, are they kind to strangers?), this bit of the brain starts to process relevant memories to figure out an answer—whether it is honest or not is not really the point. It is not used, however, when the subject is considering basic facts.

Interestingly, the medial prefrontal cortex is one of the most active parts of the brain when a person is left to do "nothing"—just thinking about whatever pops into the head. Some suggest (and some disagree) that this frontal brain region is hard at work writing the story of its life—like the time it sat in a brain scanner.

BEAM ME UP

Futurologists tell us that teleportation is "not a science problem, but an engineering problem." In other words it is possible to transmit the body from one place to another—it is just we haven't quite figured out how to do it yet. The process would involve destroying the body and rebuilding it exactly somewhere else. Would that mean you have died and then been reborn, or are you an entirely new person?

96 Functional MRI

IN 1992, THE fMRI WAS DEVELOPED AS A NEW MEANS OF SCANNING BRAIN ACTIVITY. IT REMAINS THE NEUROIMAGER OF CHOICE BECAUSE IT OFFERS A REAL-TIME VIEW OF a living brain at work.

Magnetic Resonance Imaging, MRI for short, was first used in the 1970s. It makes use of the way hydrogen atoms—including the ones incorporated into the molecules inside a human body—line up when placed in a very strong magnetic field. The orderly hydrogens in a certain region of the body can be knocked out of position by targeting radio waves at them. When they bounce back to their correct place, the atoms give out their own radio waves, and these are picked up by the scanner. Then a fiendishly clever computer program converts that signal into a image of the soft tissues inside the body.

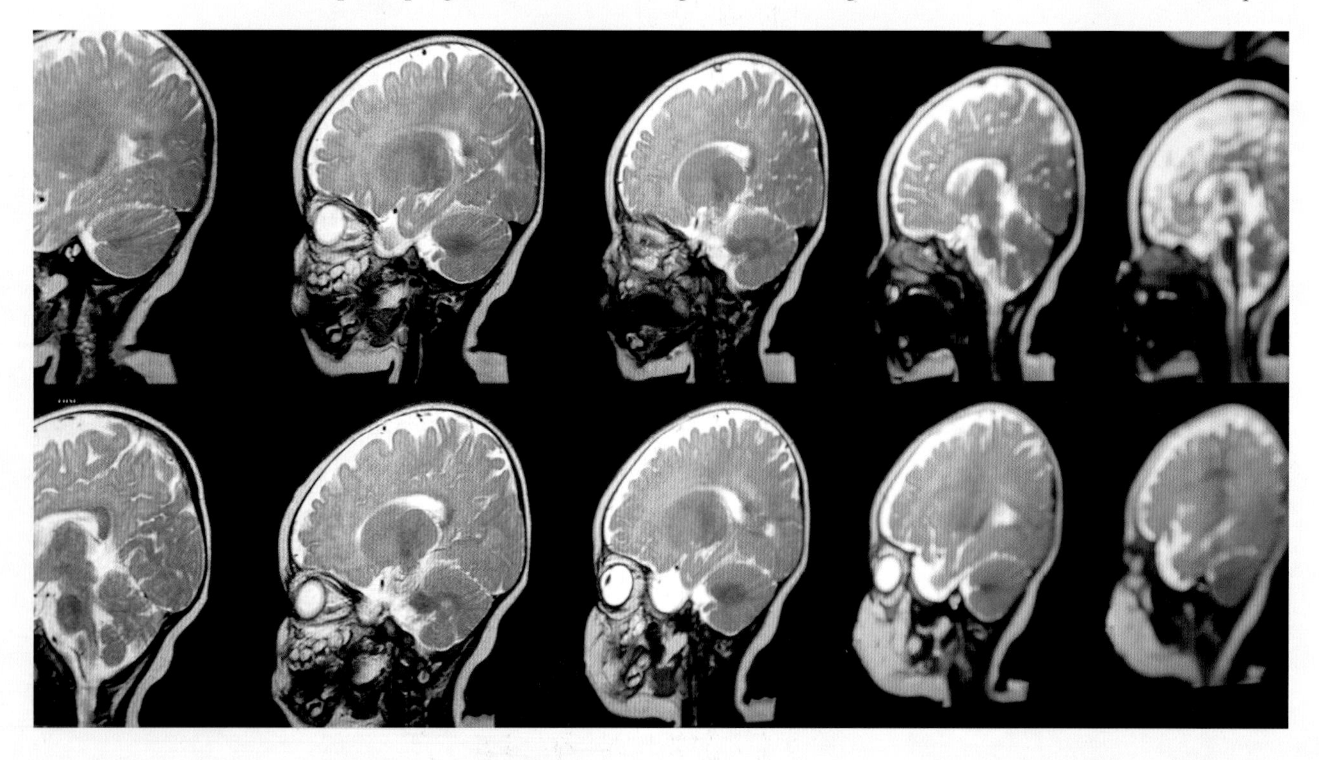

Traditional MRI scans take slices of the brain and body. However, the latest scanners are able to create a real-time video of the brain in action.

The f in fMRI stands for "functional," and refers to a new development of the system where the scanner can differentiate blood that is filled with oxygen from blood that has run out of oxygen. Brain cells do not have a store of energy, and so when they are working, the blood vessels in that region give away their oxygen. The fMRI process can highlight where this is happening, in effect, lighting up the parts of the brain that are active at that moment. An MRI machine is a supercooled magnetic tube that makes a lot of noise. However, staunch research subjects have endured many hours inside them as they are asked to perform test after test. Gradually, fMRI is helping to unpick how the dazzling patchwork of brain regions work.

97 Parapsychology

THIS IS SOMETHING OF A DIRTY WORD FOR NEUROSCIENTISTS. PARAPSYCHOLOGISTS HAVE USED TECHNIQUES THAT SEEM TO RESEMBLE THOSE USED BY BRAIN RESEARCHERS to investigate occult powers, such as telepathy and clairvoyance. Scientific scrutiny consistently debunks the results.

Reports of ghostly apparitions are as old as civilization. Until there is a better explanation, neuroscience treats them as any other delusion created entirely within the brain.

In 1994, an American neuroscientist, Michael Persinger, announced he had found a way of creating religious experiences, where people were visited by a supernatural presence. He used a device that has become known as the "God Helmet." The helmet directed magnetic fields into a person's temporal lobes. Persinger contended that the lobes worked together in creating a sense of self. When disrupted by his helmet, the left and right lobes became independent of each other, creating two identities—one perceived as the "self" and the other as a mysterious "other." The experiment was repeated ten years later by independent researchers who failed to find the same effects. Their conclusion was that Persinger's subjects were aware of the experiment's goals, and were all too willing to show that the theory was correct. (More recent research has also linked the frontal lobes with the sense of self, not the temporal ones.)

Persinger's God Helmet was a modified snowmobile helmet. The magnetic fields it produced were tiny, similar in strength to those made by a cell phone.

Science steps in

Persinger disputed this finding, but it is a common story when neuroscience meets parapsychology. Over the last century, research programs have purported to find proof of telepathy, psychic powers, and the ability to move objects with the power of the mind. When repeated under rigorous laboratory conditions, every "discovery" (so far) is shown to be flawed. One area where science seems to have discovered a link between the physical world and ethereal experiences is "infrasound"—sounds too deep for us to hear—which is known to affect mood. In 2001, a laboratory at an English university was said to be haunted: Workers in the lab reported feeling spooked by the place. Tests revealed high levels of infrasound from a faulty ventilation fan. When this was fixed, the lab became a much happier place. The speculation is that old "haunted" houses are filled with similar sounds—especially when the winds howl outside!

98 Hard Problems of Consciousness

WE HUMANS PRIDE OURSELVES ON OUR CONSCIOUSNESS. IT IS WHAT TRULY SETS US APART FROM THE REST OF NATURE, WE BELIEVE. However, no one really knows how the brain creates this awareness and without that understanding we cannot be sure what our consciousness really is.

René Descartes, the French philosopher based his famous proof of his existence, "I think, therefore I am" on doubt. Only self-aware beings doubt their own existence. However, that doubt might not be enough. Consciousness might not be all it is cracked up to be, and may even be an illusion.

Many neuroscientists seek the link between the physical brain and consciousness. This would take the form of a neural pathway or constellation of pathways that correlate to that inner voice and vision in our minds that observes events and attempts to make sense of our little corner of the world. Much of the brain's activity does not require consciousness; it can be explained without a mental process. The holy grail of consciousness research is to find which parts of the brain add the self-awareness to the other processes. However, even if this neural activity

Does this look right to you? Obviously not, since all the colors are wrong. However, your perception of color is private to you. There is no evidence to show that the sensation, or qualia, that you term blue is the same for everyone.

THE EASY PROBLEMS

There are several ways we can describe our self-awareness. These descriptions generally focus on our abilities, and frequently we do not understand how they work. Nevertheless, there is no inherent barrier to figuring them out.:

A person is conscious because:
> they are aware of the difference between wakefulness and sleep
> they can report their mental state
> they can control their behavior
> they can integrate information mentally
> they have the ability to categorize environmental stimuli and react to them in an appropriate manner
> they can focus their attention

is pinned down, it will not provide a full answer to the question of how consciousness arises and what it actually is. There will remain plenty of hard questions.

Zombie philosophy

The hard questions of consciousness were posed by the Australian philosopher David Chalmers in 1996. He used the idea of a philosophical zombie to get his point across.

Chalmer's zombie had a body but no conscious mind. When you interacted with this individual it behaved just like anyone else. If you bumped into each other, knocking heads perhaps, both you and the zombie would rub your heads, say "Ow!" and then apologize for hurting the other. You feel your pain and assume that the other person feels the same kind of thing. However, you have no evidence of that. In fact the zombie has no mental process and is an automaton simply following a set of rules that leads to the same kinds of behavior as yours. Chalmers used this thought experiment to question whether we can ever verify that the internal sensations and representations that make up our consciousness are the same as those experienced by others. These private qualities—such as color, smell, and the sensation of pain—are termed qualia. We all agree that the sky is blue but that does not mean the qualia for blue is always the same in every consciousness. If we could climb into someone else's mind, we might find their qualia for blue is actually our pink one. And the zombie's brain can detect the color blue, it can point to the sky, and say, "It is blue!" but it has no qualia for blue. However, one piece of research suggests that having qualia does not automatically disqualify us from being zombies!

"I think the existence of zombies would contradict certain laws of nature in our world. It seems to be a law of nature, that when you get a brain of a certain character, you get consciousness going along with it."

DAVID CHALMERS

READY, GO!

The readiness potential is a tiny reduction in voltage in the areas of the brain associated with movements. It appears before a voluntary movement is actually commanded by the brain. Studies suggest that the brain prepares to make a movement before we have consciously decided to do so. So what told it to get ready?

Getting ready

In the 1980s, researchers were able to measure the relationships between a conscious thought and its physical outcome. To do this they made use of the so-called readiness potential (*see* box). The research found that the readiness potential appeared a fraction of a microsecond before the human subjects made the decision to move! So the brain appears to be acting on a decision before it is made. This has three possible implications. First, our consciousness has a time delay. We are only aware of what we are thinking a fraction of a second later—but that awareness is a complete representation of the cognitive process. Or second, consciousness is an illusion, an overlay to give sense to our actions but has no actual influence over them. That would mean we have no free will and are under the control of an unconscious process. The third possibility is that our actions are initiated by the unconscious mind, and our consciousness is there simply to veto them. Instead of free will, this concept is sometime described as the power of "free won't."

99 Personality or Neurology?

Where does brain anatomy end and personal responsibility begin? This question is increasingly being asked in criminal trials where defense teams argue that the accused is not to blame. It was their brain that did it.

In 2002, a school teacher from Virginia was convicted of a crime against children. The night before he was due to start his jail term, he went to the doctor complaining of a crippling headache and confessed to an unendurable urge to commit further crimes. He was unable to control himself in a proper way. Ensuing medical examinations revealed he had a brain tumor putting pressure on his frontal cortex. This area is associated with impulse control and judgement. Once the tumor was removed, the man ceased to have the same criminal urges. Doctors testified that the tumor was responsible for the marked changes in the teacher's personality that had led to his crime and conviction. He was released from prison.

Impulse control

This was an unusual case, where a man of previous good character had turned into a danger to others. However, scientists began to wonder if other criminals had some deficiency in the frontal lobes that made it hard—if not impossible—to control their behavior. Since 2002, the number of trials in the United States involving neurological evidence has climbed steadily. Within ten years more than 1,500 serious crime trials (probably more) used neuroscience to mitigate a convicted felon's actions. In some jurisdictions it is becoming common for lawyers to brainscan all their clients. Critics of this approach argue that the division between a competent brain and an incompetent one, where the owner is not criminally responsible, has yet to be drawn. Neuroscience is built on a biological basis for the brain's functions, so could it not be said that everyone's actions—including those of the judge and jury—are beyond their ultimate control?

PERSONALITY DISORDERS

Some people are sick just because of their personality. There are around a dozen personality disorders, including:

Paranoid	**Suspicious and mistrustful**
Borderline	**Unstable and impulsive**
Narcissistic	**Grandiose self-image**
Obsessive-compulsive	**A rigid conformity to rules**
Avoidant	**Socially inhibited and easily humiliated**

The famous and the infamous are remembered for their achievements, but do we thank (or blame) them or their brains? It is said that Adolf Hitler could fill a psychiatry text book all by himself; Greta Garbo wanted to be alone because she had an avoidant personality disorder; Henry Ford's automobile empire was a product of his narcissistic personality disorder (he ignored any innovations that were not his); while Vincent van Gogh may have had borderline personality disorder.

100 Computer Brains

A MACHINE THAT IS INTELLIGENT AS A HUMAN IS ONE THAT HAS BEEN A LONG TIME COMING. The goal of artificial intelligence (AI) was set in 1955, and so far we are still waiting. Is an electronic device capable of replicating the functions of a biological brain? Or will AI beings require a new approach?

"AI would take off on its own, and re-design itself at an ever increasing rate."

STEPHEN HAWKING

In some ways a brain and a computer are similar. Both require a supply of energy, both send signals in the form of electricity, and both have a memory that can be modified. But that is really where the similarity ends. Computers lead when it comes to storage and processing power. The fastest supercomputers can hold 30 million gigabytes to the brain's 3.5 million—this is a good guess, not a solid figure. The machines are also faster, performing 8,200 trillion operations in a second compared to the human brain's 2,200 trillion. However, the human brain is the model of efficiency. It relies on a "program" of 750 megabytes (that's the amount of information in the human genome) and 20 watts of power compared to the supercomputer's 10 megawatts. Also the computer fills a room, while the brain ... well it fits inside your head.

However, a brain does not function like a digital computer. If it did then the supercomputers would be very intelligent indeed, when in fact they are so dumb they can't do anything unless we tell them to. The difference is that the brain appears to operate multiple parallel processes at once, something an electronic computer cannot do. If computers are to become truly intelligent they will need to copy the brain's system—only no one really knows what that is.

BEWARE AI

In 2014, Stephen Hawking, a world-leading physicist, warned of the dangers of creating artificial intelligence (AI). Unencumbered by biology, he said, AI could direct its own development at a rate beyond our comprehension—and it would not have our best interests at heart.

NEURAL NETS

You may have heard that AI computers run using a neural network. In terms of computing, this is a processor that is designed to connect itself together, rather than follow a preset circuitry like the microchips in everyday devices. This system uses multiple layers of nodes—the computing equivalent of a neuron. The top layer receives inputs (stimuli) and the bottom one produces outputs, but what that output is depends on how the information has traveled through the intermediary layers. This smart computer tries out all possible connections and tests the resulting output. It learns from its mistakes which output is the right one for each input.

The idea that AI computers will be electronic versions of our brains is falling from favor. Instead, an entirely new form of computing will be required to create AI.

101

The Brain: **the basics**

IT IS VERY HARD TO PINPOINT ANYTHING ABOUT THE BRAIN THAT IS BASIC. However, let's have a look at what all this study adds up to, by exploring how the brain and body connect and interact. First, we will look at the way the brain handles its touch senses and the muscles that respond. Then we will follow the different nervous pathways that run around the body.

Touch and movement

The somatosensory cortex and its neighbor, the motor cortex, were the first major control centers mapped on to the brain by neuroscientists. The former is concerned with the touch senses. There are several kinds of stimuli that create inputs to the system, such as heat, different levels of pressure, and pricks, which are all processed here. The motor cortex initiates voluntary movements, perhaps in response to the information arriving at the cortex next door.

It is possible to present the way different regions of the brain represents, the body as *homunculi*, or "little people." These are oddly shaped bodies, where the regular dimensions have been changed to reflect how much of the brain is devoted to different parts of the body. As you can see, the sensory homunculus devotes more brain to picking up touch information from the fingers and hand than it does for the entire torso and legs. The fig leaf needed to secure the modesty of the sensory homunculus is considerably larger than the one on the motor homunculus, who is an altogether more grotesque figure. This shows that moving the lips and tongue requires a lot more brain space than the rest of the face, while the dextrous hands dominate much of the cortex. Take a good look. You'll never see yourself in the same way again.

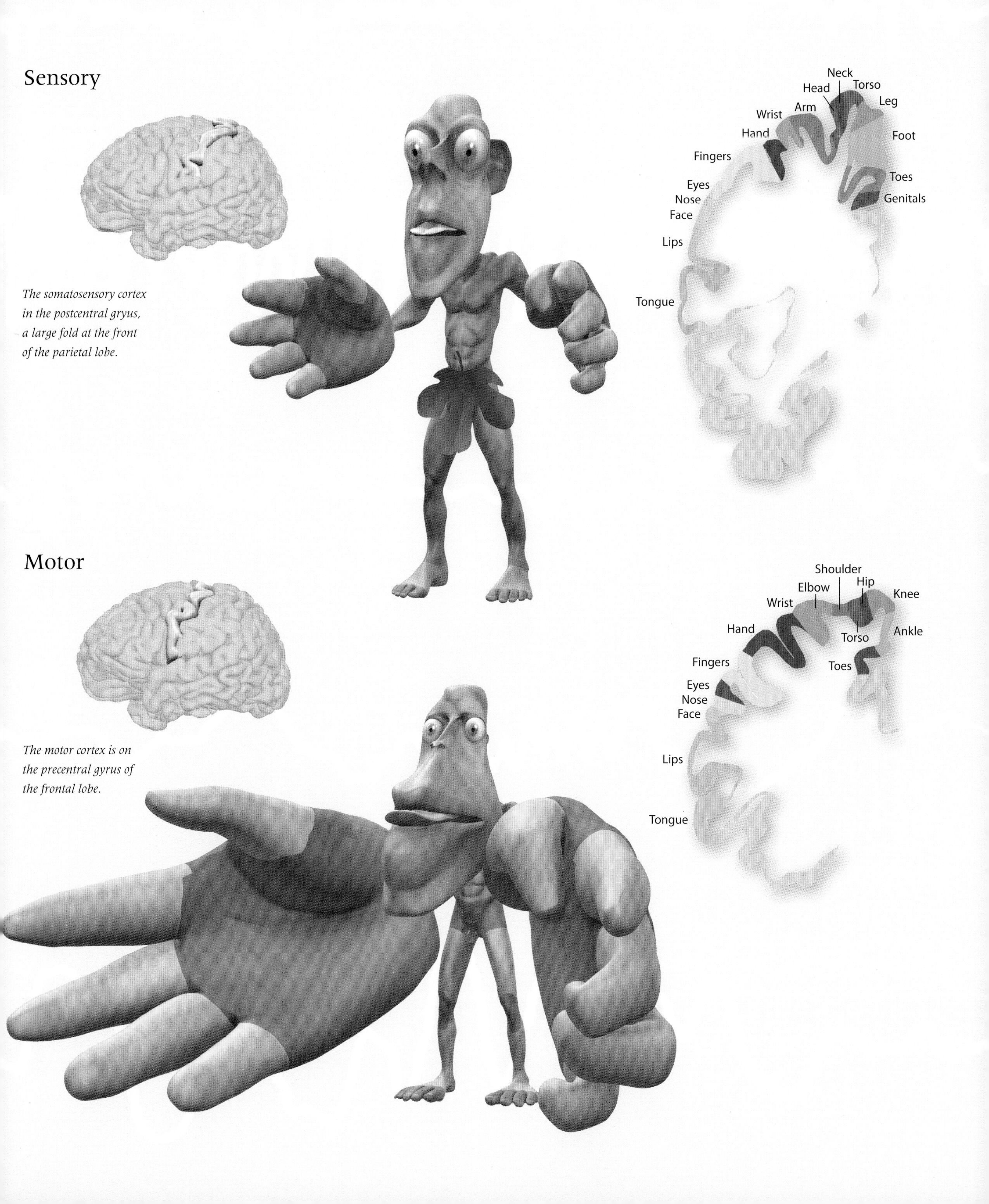

The somatosensory cortex in the postcentral gryus, a large fold at the front of the parietal lobe.

The motor cortex is on the precentral gyrus of the frontal lobe.

Major nerves

The brain is the main feature of the central nervous system. The body has another system: The peripheral nervous system which is a network of nerve fibers that connect to all parts of the body largely, but not exclusively, through the spinal cord.

Body pain and other sensations and voluntary movements are carried via nerves and the spinal cord.

Spinal cord

The spinal cord is the lower part of the central nervous system. It is a bundle of nerve fibers that connects the brain to the peripheral system. The cord is divided into four sections: The cervical, thoracic, lumbar, and sacral divisions. There are 30 nerves (31 if you count the tailbone nerve), and their position down the spine relates roughly to that region of the body.

Somatic nerves

The somatic nerves are the ones that carry the touch sensory information to the brain and then send out the commands to the muscles for voluntary movements. Involuntary reflex movements also use this system, only the signals do not get to the brain.

Cranial nerves

Some peripheral nerves come straight out of the brain. Most of these are associated with the head and neck, while the vagus nerve descends to the viscera where it is involved in another part of the peripheral system: the autonomic nervous system. Let's look at that next.

Somatic nerves

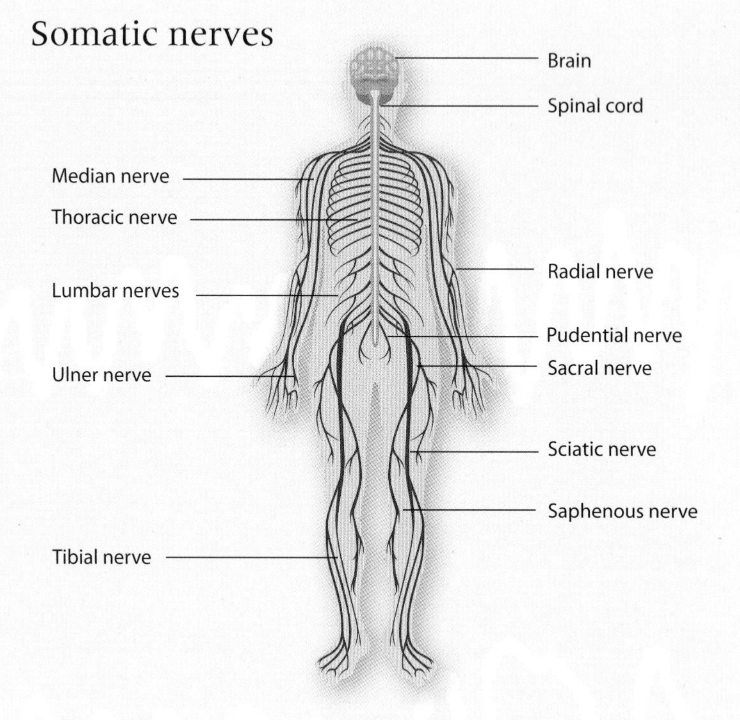

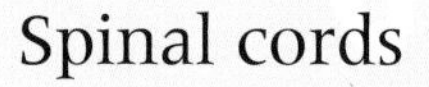

Spinal cords

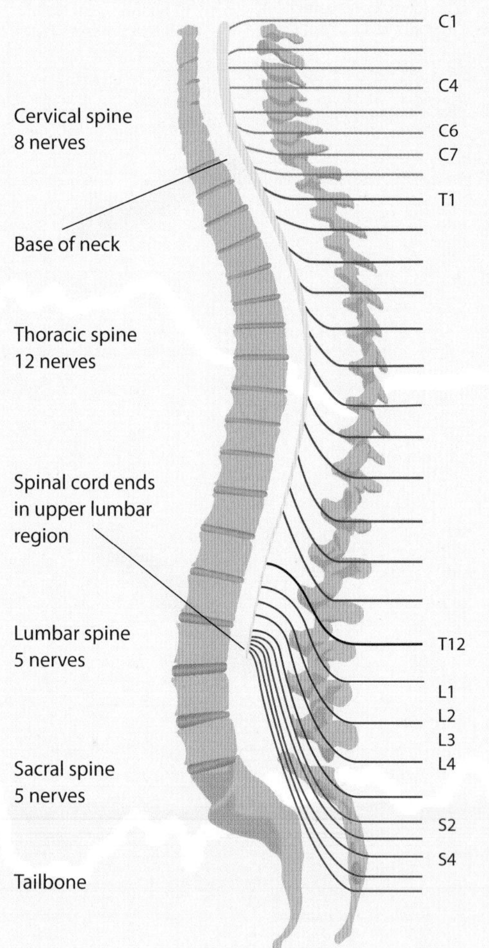

Cranial nerves

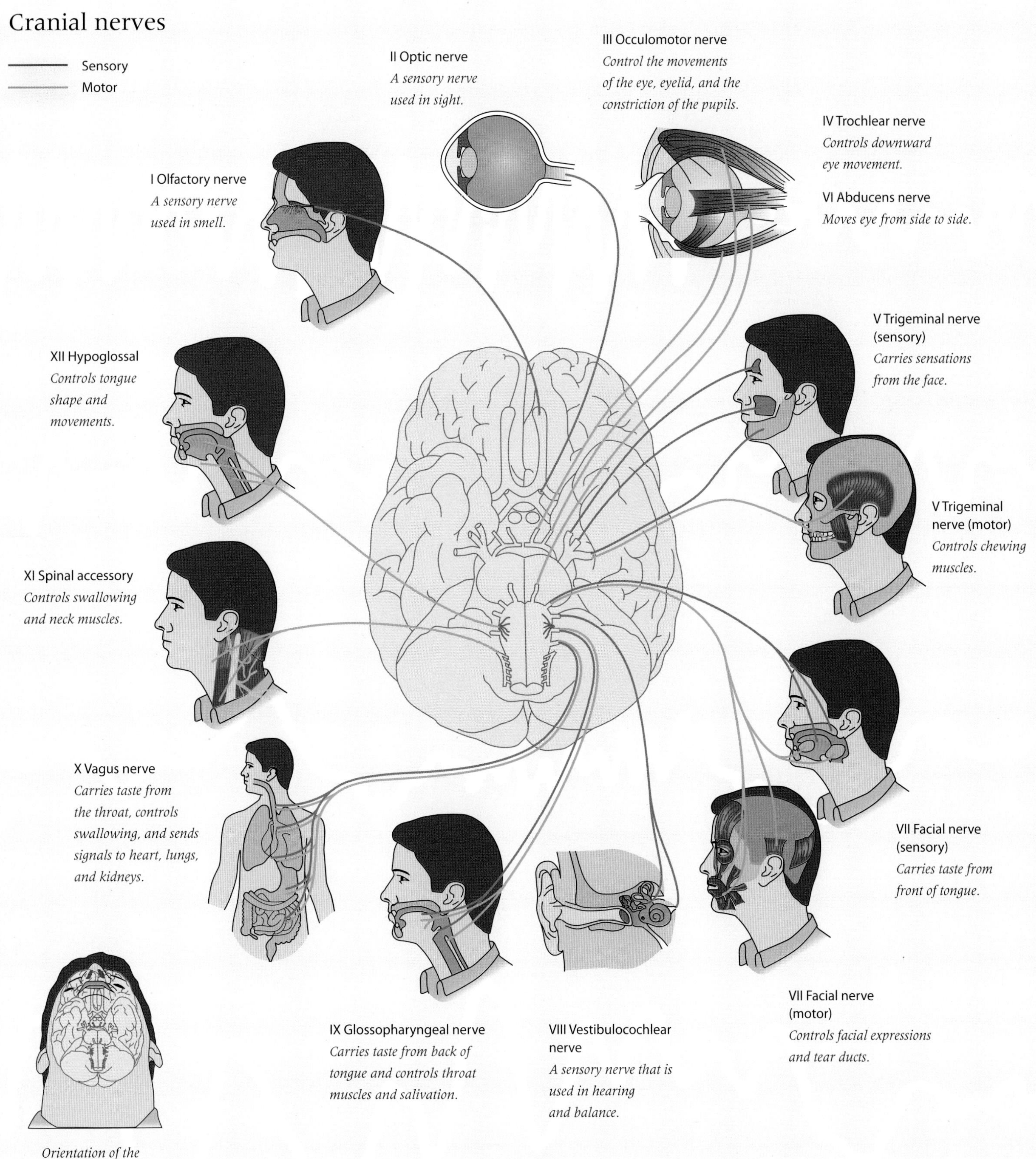

Orientation of the main brain diagram.

The autonomic nervous system

Many processes in the body appear to be automatic and self-regulating, However, they are still under nervous control through the autonomic nervous system. This is in two parts: The sympathetic system and the parasympathetic system. Both are antagonistic to each other, meaning they work to influence the body in opposite directions.

The size of the iris is controlled by competing instructions from the sympathetic and parasympathetic system.

Sympathetic system

This side of the system controls the fight or flight response. It galvanizes the body to prepare for action. It does this stimulating the adrenal gland on top of the kidney. This makes the hormone epinephrine (also known as adrenaline) flood into the bloodstream. That causes blood to flood to the muscles so they are ready to work. Alongside that, the sympathetic system is slowing down digestion, relaxing the lungs so they can take in more air. The system also controls other systems not required for fighting but they get involved, too: The salivary glands stop working and the mouth dries out; the tear ducts stop working—you don't cry when you are scared; and the bladder can relax completely.

Parasympathethic

The parasympathetic system restores the body to a relaxed, normal state—allowing it to "rest and digest."

Lungs
Sympathetic: Relaxes airways
Parasympathetic: Constricts airways

Heart
Sympathetic: Speeds up beating rate
Parasympathetic: Slows heartbeat

Eye
Sympathetic: Dilates pupils
Parasympathetic: Constricts pupils

Stomach and pancreas
Sympathetic: Inhibits digestion
Parasympathetic: Stimulates digestion

Liver
Sympathetic: Stimulates glucose production
Parasympathetic: Stimulates gallbladder

Bladder
Sympathetic: Relaxes bladder
Parasympathetic: Contracts bladder

Tear gland
Sympathetic: no link
Parasympathetic: Stimulates tears

Salivary glands
Sympathetic: Inhibits salivation
Parasympathetic: Stimulates salivation

Adrenal gland
Sympathetic: Stimulates release of epinephrine (adrenaline)
Parasympathetic: No link

Intestines
Sympathetic: No link
Parasympathetic: Dilates intestinal blood vessels.

Rectum
Sympathetic: No link
Parasympathetic: Constricts sphincter

Genitals
Sympathetic: Stimulates orgasm
Parasympathetic: Stimulates erection and arousal

Sympathetic

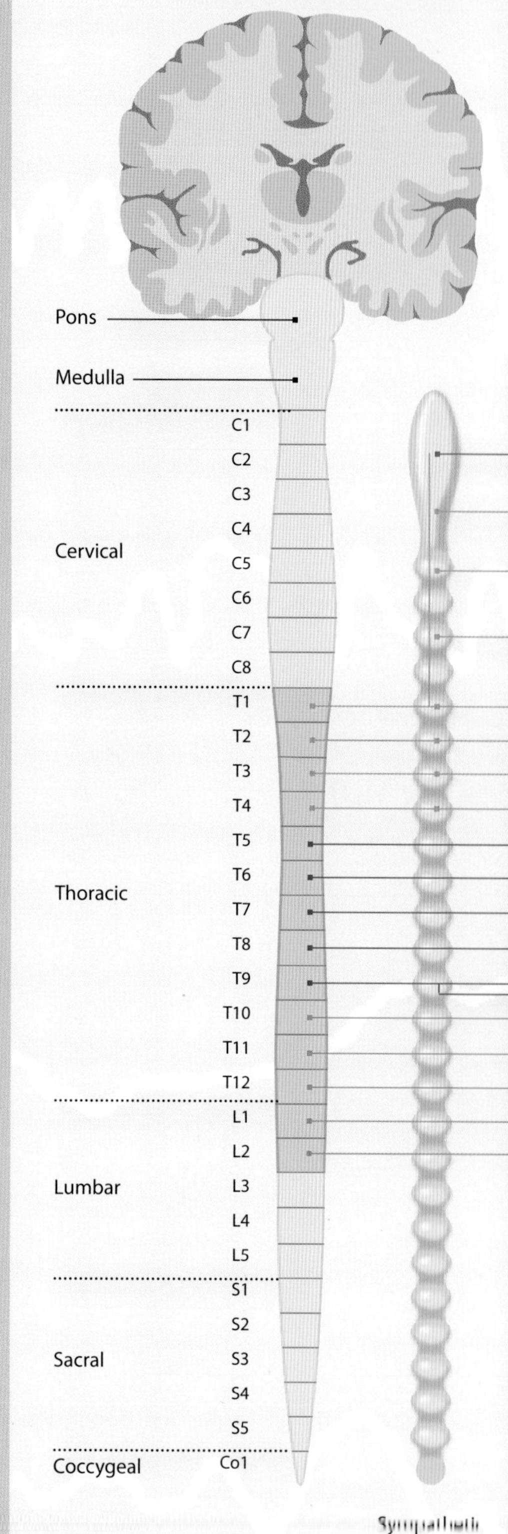

Parasympathetic

The sympathetic system runs through two chain ganglia that run either side of the spinal cord.

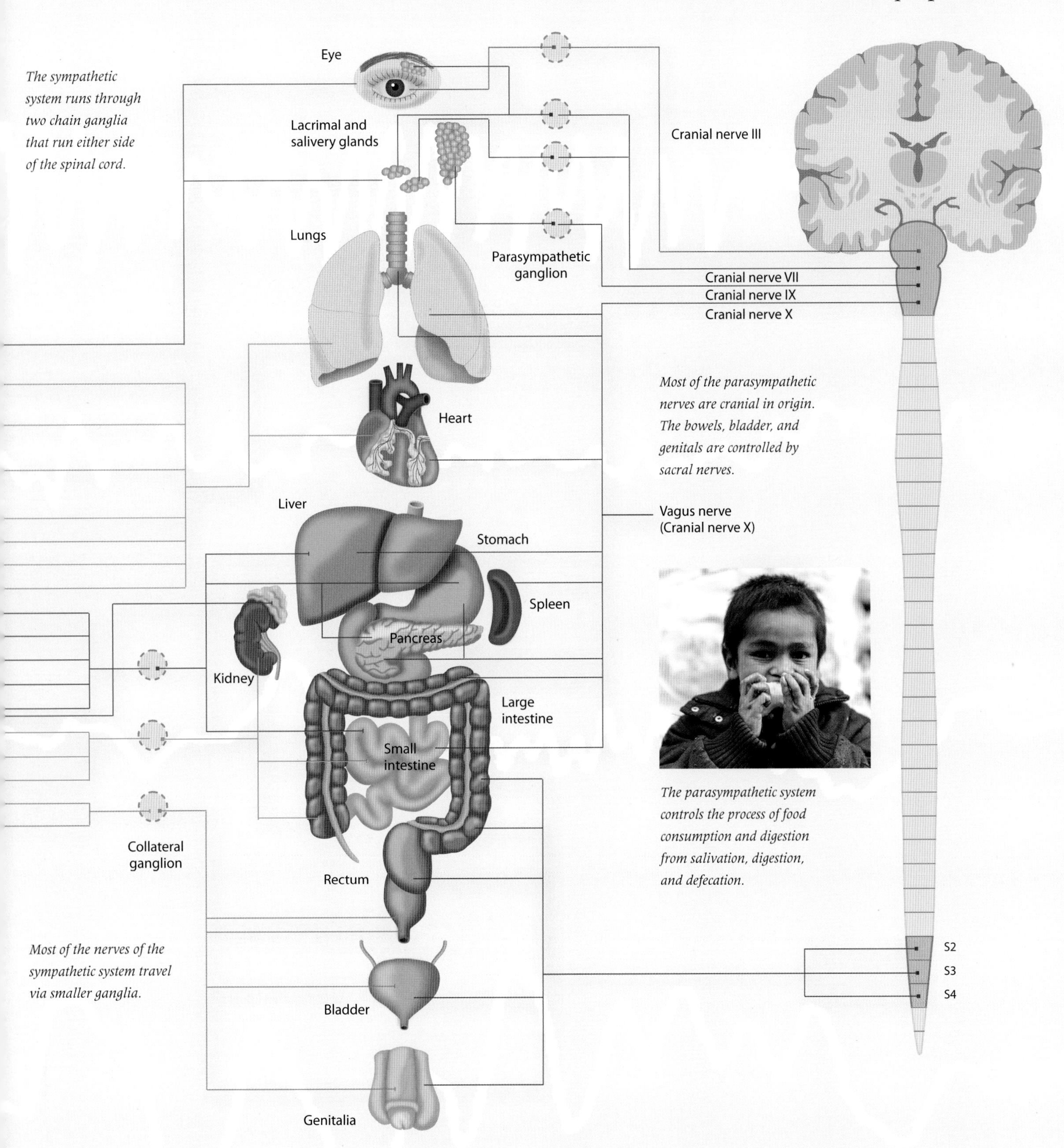

Most of the parasympathetic nerves are cranial in origin. The bowels, bladder, and genitals are controlled by sacral nerves.

The parasympathetic system controls the process of food consumption and digestion from salivation, digestion, and defecation.

Most of the nerves of the sympathetic system travel via smaller ganglia.

THERE HAVE BEEN GREAT LEAPS IN NEUROSCIENCE IN RECENT YEARS, AND OUR UNDERSTANDING OF THE BRAIN TODAY IS INCOMPARABLE WITH JUST 50 YEARS AGO. However, brain science is a new field and there is much still to discover. Let's take a look at a few questions needing answers.

Why are most of us right-handed?

Around 85 percent of humans are right-handed. We generally associate this concept with using the right hand to hold a pen as we write. However, literacy is not yet universal, and a couple of hundred years ago the majority of people could not write—but they were still right or left handed. Instead of putting pen to paper, they were using their dominant hand for eating and using tools, and—most importantly—making things.

However, we are getting ahead of ourselves. To be "handed," we first need hands, and to get those our ancestors had to walk on two legs, thus freeing up the front limbs for manipulating objects. Were the earliest hominids handed? The best evidence we have comes from apes. When they are walking on all four legs, great apes do not exhibit a dominant hand. However, when moving upright, each one does favor a hand—with a 50:50 split between left and right. Anthropologists made use of their own handedness to take the investigation further: Left-handed and right-handed scientists made stone tools using the same methods as early humans. They compared their results to ancient artifacts unearthed from prehistoric times. The results showed that right handedness only became dominant over the left side about 600,000 years ago. What had changed? It is well documented that animal brains divide their labors between both sides. For example fish are more likely to spot food using their right eye (and the left side of their brain). It is suggested the right brain is doing something else—perhaps looking out for threats. So our four-legged ancestors probably already had brains that divided tasks between one side and the other. The marked shift toward handedness that began 600,000 years ago is thought to be nothing but an accident of nature. Most human brains use the left hemisphere to process communicative language—the meaningful stuff, not just animal grunting. The leading theory is that as language evolved on the left, the right side of the body (which is controlled by the left brain) also became dominant, most notably the right hand. But to prove it we'll need to brainscan a cave man. Hands up who thinks that will happen.

Nearly all of us rely on one hand over the other. Could that just be an accidental consequence of language?

What is the point of crying?

Are you crying, or is there just something in your eye?

Only humans cry. Well, only we cry for emotional reasons. Every mammal eye can produce tears, which have a utilitarian cleaning function. A crying human tells us something about ourselves. The tears function as a signal, and we deploy our theory of mind to understand how the crying person must be feeling—we remember how we feel when we cry. In that way, crying is one of many facial expressions humans use to communicate, and the message is one of extreme, deeply felt emotion, something beyond words, at least not right there and then. Another theory points to the hormones present in tears. It suggests that the reason why tears became a signal of emotion in the first place was because they were initially a method of excreting the chemicals produced by the pituitary gland when the body is under stress. However it arose, the system is a powerful one—we humans can cry at the death of a pet animal, while the animal's biological family do not shed a single tear.

Is the brain clever enough to understand itself?

This is something of a paradox. We only have the human brain with which to understand how the human brain works. As far as we know the human brain is the most complex system in the Universe. However, if we are to understand it we require it to follow some kind of simple rules—simple enough, that is. However, if the brain system is simple, then will it be clever enough to figure out the way it works? So far the answer is that the brain has not been clever enough to understand itself. However, since records began about 100 years ago, humans appear to have got steadily cleverer. One assumes the brain's complexity is the same now as it was 100 years ago, and we (individually and collectively) are just getting better at using our gray matter.

This clever photograph was produced by human intelligence. Are we clever enough to figure out what that is?

IMPONDERABLES

What is the minimum that the human brain does?

There are two ways of considering this question. The first is through tragedy and points to the minimum the brain does for a person to still be regarded as alive. The answer lies in the hind brain, where the respiratory center can still induce regular breathing while the rest of the brain does nothing at all. According to the rules, this brain is not dead, although it will never recover. However, a more fun way of tackling the question relates to the oft-quoted "fact" that regular folks only use ten percent of the brain, while the odd genius is able to harness more of it to reach new levels of enlightenment in some field or other. This is bunkum and stems from a flippant comment attributed to Albert Einstein, who was trying to elucidate how he had made so many discoveries, seemingly all at once. We use all of our brain, although never all at once. (The brain is nevertheless very demanding on the rest of the body. A fifth of the oxygen entering the lungs is used simply to keep the brain ticking over.) The brain does have a "resting state," a baseline of alpha waves at which it could be said that it is doing "nothing." Of course it is still doing a great deal but researchers seek this as a baseline to be able to detect the elevations in activity that show the brain is doing "something." The resting state is a pattern of activity that can be seen in all brains. And the questions remain. The resting state is largely unconscious activity that continues even when we are doing nothing. When the brain is doing "something"—responding to a stimulus—how much of that activity is still part of the unconscious? Is "consciousness" only associated with the elevated portion of activity?

How does the brain record time?

Even without a clock we know that time is passing. Nature gives us an obvious cue with the rising and setting of the Sun, and our bodies follow suit with the rhythms of hunger and sleep that punctuate our day and night. This is the circadian rhythm, ultimately calibrated by changes in light levels. However, the brain needs to perceive shorter time periods. Playing music or sport or simply speaking requires the brain to coordinate movements according to rhythms measured by the millisecond. It is assumed this kind of timekeeping is controlled in the cerebellum, but no one really knows. Also, differentiating periods of a few minutes or hours is crucial for planning activities and ordering memories. You may repeat the same task several times during the day, but the brain gives each one a unique place in time. The best theories of how the brain does this involve the hippocampus. This is believed to be the video recorder of the brain, able to record events, and then play them back in an edited form that may or may not make it into long-term memories.

Does the brain have a clock, or several clocks? Are they in time with each other?

When you forget something is the memory gone or just the ability to recall it?

The memory trace is still too faint for neuroscience to see properly. Neuroscientists believe that the brain may organize memories according to different associations, but where a memory physically resides in the brain has alluded them so far. The best they can do is observe the brain activity associated with the act of recalling a specific memory and represent that as a cluster of three-dimensional pixels, called "voxels." Scanners cannot see closely enough to translate a voxel of activity into a specific network of neurons, but the system is good enough to throw light on how we recall memories. Recent research suggests that memories from our past are in a constant battle to be remembered. In order to recall the voxels of one memory, the brain has to forget, or at least weaken, the voxels of another. In other words the brain actively forgets memories. This translates into real world scenarios: If the witness of an event is quizzed about it and asked to recall the story of what happened, the salient memories become reinforced and vivid, while others are repeatedly suppressed. This might be why during cross examination in a trial, a witness's memory can be shown to be vague about certain aspects of an event. For example, the witness will struggle to remember what they had to eat for lunch that day or what they were wearing. Perhaps some would think that vagueness indicates they are lying, others might point to the way that remembering something also involves forgetting something else.

Henri Rousseau's The Dream *from 1910. It is art but is it really of a dream?*

Why do we dream?

If you think anyone knows for sure, wake up, you're dreaming! There are two schools of thought about dreaming: A physiological take on the subject is that the dream is an artifact, an accidental—and meaningless—feature of the sleeping brain's activity. Physiological arguments suggest that our consciousness is activated during REM sleep as the brain processes the lessons of the day. Any meaning we attribute to this consciousness is pure fantasy, testament to our abilities to fit patterns to things. The psychoanalytic approach suggests that a dream is a conscious projection of the activities of the subconscious. This theory—largely discredited nowadays—would suggest that the content of the dream reflects the things that are most concerning to us, and provide a window onto the unseen battle of ideas that takes place subconsciously. To really understand why we dream, we need to understand why we sleep. And again no one really knows. While sleep provides a rest and healing period for the body, the brain remains largely active during sleep, leading some to suggest it is "doing the admin." While we sleep the brain may be processing the memories of the day, dispensing with some, storing others, and even performing some executive-level thinking—sleeping on a problem always helps, they say.

IMPONDERABLES

What code do nerves use?

Many of us are familiar with the way a computer is programmed using machine code, a string of 1s and 0s. These numbers represent variables, bits of data, user inputs, and machine outputs, but they also represent the instructions that control how the processor handles all those variables. How that all works is perhaps academic here, but it frames a question about the brain: Does that function using some kind of "brain code?" The means by which the nerve impulses move along neurons and then on to the neighboring ones is now well documented. Also known is that a signal from one "upstream" neuron can either stimulate the "downstream" neighbor, causing it to send a signal onward, or inhibit it and stop this or other signals from upstream being passed on. It used to be thought that variables were coded by the frequency of neuron signals. Frequent pulses have more of an effect than infrequent ones. However, it is now believed that neurons may modulate the firing rate. The cell alters the time between each successive electrical pulse by a few milliseconds, as a means of encoding information. As well as understanding the brain's hardware, there is still much to learn about the software as well.

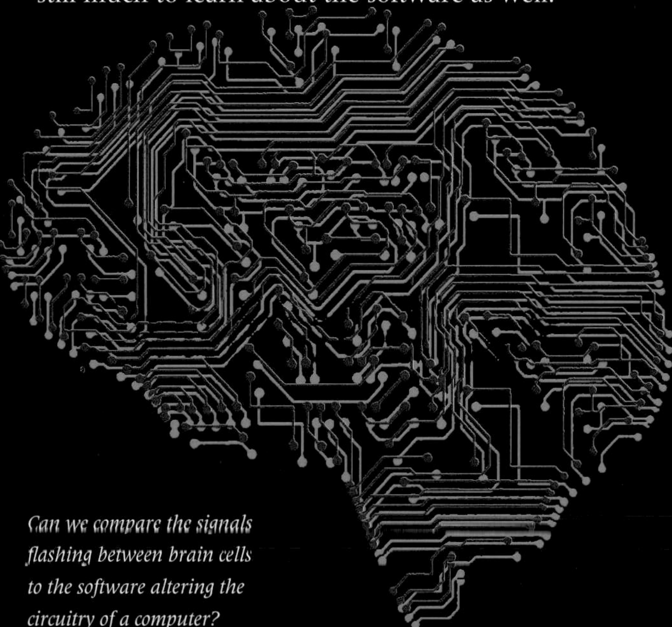

Can we compare the signals flashing between brain cells to the software altering the circuitry of a computer?

Is it possible to preserve a brain after death—and then restart it?

Medical science is pushing back the moment of death, second by second, minute by minute. It is not unusual for people to be without signs of life for four or five minutes and then be resuscitated to go on to lead normal lives. There are ways in which that time period can be extended. However, the central problem remains the same: It is generally understood that a few minutes after the heart has stopped, and with it the supply of oxygen in the blood, the brain (and the rest of the body) has died. This is not really true. The cells will begin to die but it takes much longer than five minutes. However, if the heart is restarted after that kind of time period, the reperfusion of the blood—a sudden return of the blood pressure—will destroy much of the brain's tissue. So as things stand today, the body has gone beyond a state that can support life—any attempts to do so simply hasten its destruction. If injuries from reperfusion can be solved then the time of death could be delayed further still, perhaps by hours. Some people have chosen to stop the clock at the moment of death and freeze their brain or entire body in liquid nitrogen. If that person has died without damaging the brain—i.e. there is no brain death prior to cardiac death—then it might be possible in future to thaw them out and reperfuse their body with no ill effect. However, there are still huge obstacles to this process. The one thing these frozen folks have on their side is time.

Our brain's perception of the world is a mental model based on what came before.

How does the brain predict things?

The brain is a prediction machine. The way we understand the world is to model the likely outcomes of events based on our past experience, and constantly update it as time moves forward. This is how we are able to understand speech. We are not comparing sounds against a database of meanings; we are predicting which sounds are likely to follow the next based on our own fluency with the language. In so doing we are aware of what the speaker might say next before he or she actually does. This ability to predict the near future is central to our human form of intelligence. However, the location of the "prediction center" in the brain (should such a thing exist) and how it might actually work has so far alluded neuroscience.

How do we learn to talk?

Implicit in this question is the "poverty of stimulus" concept. Some linguists have asserted that language is too complicated for infants to learn so quickly given the relatively low exposure they get to the language's rules. A baby listens to others and begins to talk back. After communicating by repeating words, it then begins to use sentences. The child is not repeating these. It is able to construct correct sentences that it will never have heard another person say before. These linguists say that the child has not been given enough opportunity to figure out the language rules by this stage. Therefore, they say, there must be some innate language center in the brain that allows them to acquire languages according to preset rules. To back this up they point to the way that languages appear to share similar syntax. However, these ideas are now rather old-fashioned. Since the 1960s, when they were proposed, our understanding of the true power of the brain—and indeed of artificial intelligence—has helped to turn most people against the argument. One alternative is that the infant brain applies pattern recognition to the words and sentences it hears, and figures out the rules behind language that way. After 18 years of learning, the average English-speaker knows 60,000 different words (although seldom uses most of them). The next question is how can we be sure we know what any of them mean? The only way is to ask someone else—but how do they know?

As I was saying...

The Great Neuroscientists

ALTHOUGH THE TERM NEUROSCIENCE HAILS FROM THE LAST FEW DECADES OF THE 19TH CENTURY, great thinkers have been focused on the form and function of the brain for millennia. Here we take a look at the lives behind the discoveries that have advanced our understanding of the brain through the centuries. Pushing back the boundaries of neuroscience involved a mix of laboratory experiments and clinical observation. Our great neuroscientists are mostly physicians and surgeons but they all had their own unique approach to figuring out a small piece of the incredible puzzle that is the human brain.

Herophilos

Born	c.335 BCE
Birthplace	Chalcedon (now Turkey)
Died	c.280 BCE
Importance	Founding figure of anatomy

Agreed by most as the first anatomist, this Greek doctor spent much of his working life in Alexandria. Away from the Greek heartland in the upstart colony at the mouth of the Nile, Herophilos was free of the taboo on dissecting human bodies, so much so he often did them in public. Some reports say that he also cut into several hundred living convicts as well! After Herophilos's death human dissection became frowned upon again and was not revived for many centuries to come. Herophilos recorded his discoveries, which included the difference between arteries and veins, in nine books, all of which have now been lost.

Plato

Born	c.428 BCE
Birthplace	Athens, Greece
Died	c.348 BCE
Importance	Proposed head is seat of intellect

Plato was from an aristocratic family. As a young man he was taught by Socrates, and most of what we know of that great Athenian philosopher is from his pupil's accounts. After Socrates's execution, Plato founded a school called the Akademia, which has had a lasting impact on Western thought. The name probably derives from the previous owner of the land. One of its famous pupils was Aristotle. Plato's given name was Aristocles, but given his link to Aristotle, it is perhaps fortunate for history that he was nicknamed Plato, meaning "broad," by his wrestling teacher.

Galen

Born	c.130 CE
Birthplace	Pergamon (now Turkey)
Died	c.216 CE
Importance	Leading figure in medicine

As a young, wealthy man Galen traveled around the ancient world to absorb as much medical knowledge as possible. He did his initial training at home, in the coastal city of Pergamon, and toured the Greek islands (the home of Hippocrates) and eventually went to the medical school in Alexandria. The skills he accrued made him the leading doctor back home, where he tended to the local gladiators. In his late 30s, Galen moved to Rome, and became well known for his treatments during a plague that hit the city in 166. His fame led to him being the personal physician to two emperors. He lived to an old age, but his final years and date of death are disputed.

Alhazen

Born	c.965
Birthplace	Basra, Iraq
Died	c.1040
Importance	Proposed optical theory of vision

In Europe of the Middle Ages, al-Haytham was known simply as "The Physicist." He is perhaps the most prolific of the scientists from Islam's Golden Age. His hometown of Basra was a cultural hub in the 10th century, but al-Haytham finished his education at Baghdad's House of Wisdom, the top academic institution of the time. However, al-Haytham was not that wise. The story goes that his ill-fated move to Cairo was down to him boasting in Baghdad that he could control the Nile—a boast that landed him in deep water.

Thomas Aquinas

Born	January 28, 1225
Birthplace	Roccasecca, Italy
Died	March 7, 1274
Importance	Developed localization of brain functions

Thomas Aquinas belonged to a minor lineage of a noble family. While his brothers looked for advancement in the military, Thomas opted to join the Dominicans, at the time a newly formed Christian order with a reputation for intellect. Aquinas studied under Albertus Magnus in Cologne, Germany, who proved to be a lasting influence, and then went on to have two stints as master of theology at the Paris university (one of the first in Europe). In 1273, he suffered a seizure, which was interpreted as a vision, which made him renounce his life's works as "mere straw." He died the year after.

Avicenna

Born	c. August 16, 980
Birthplace	Bukhara, Uzbekistan
Died	December 10, 1037
Importance	Islamic physician and scholar

Known as Ibn Sina in Arabic, Avicenna proved to be a bright spark from an early age, outwitting his teachers in science and philosophy. The story goes that he administered to the health needs of the local emir while still in his teens. One benefit of this position was that Avicenna was given access to the royal library, allowing him to spread his sphere of interest to encompass physics and politics, accruing considerable wealth in the process. In the end he wrote 200 books himself, and would have produced more if he had not been poisoned by one of his servants.

Andreas Vesalius

Born	December 31, 1514
Birthplace	Brussels (now Belgium)
Died	October 15, 1564
Importance	First modern anatomist

Andreas Vesalius was so good at dissections that as soon as he graduated as a doctor from the University of Padua, they offered him a job as head of department. His name is a Latinization of Andries van Wesel, which betrays his Dutch roots, despite spending his working life in Italy and ending up as the court physician for the Holy Roman Emperor, Charles V. We elevate Vesalius's achievements in anatomy thanks in the main to the exquisite illustrations that display them. However, Vesalius did not draw them. It was Johan van Calcar, a Venetian artist did the work that has allowed Vesalius's name to persist for so long.

Johann Jakob Wepfer

Born	December 23, 1620
Birthplace	Schaffhausen, Switzerland
Died	January 26, 1695
Importance	Discovered cause of stroke

Wepfer is recognized as the first person to link strokes with a problem with the blood supply in the brain. This achievement is celebrated by the annual Wepfer Award given for contributions to stroke treatments. In life, Webfer was much in demand as a physician, tending to many of the royal families that ruled over patches of Central Europe at the time. Wepfer was a renowned anatomist, but also had interests in the study of natural poisons, such as hemlock, and was one of the first to point out that using elements such as arsenic and mercury—a very common practice in those days—caused more harm than good.

Giovanni Aldini

Born	April 10, 1762
Birthplace	Bologna, Italy
Died	January 17, 1834
Importance	Studied the role of electricity in bodies

It could be said that Giovanni Aldini took up the family business. His uncle Luigi Galvani had become famous for his discovery of electrical currents running through animal tissue. While other researchers used this discovery to build electrical technology, Aldini became famous for his public demonstrations. He showed how electrical current could be used to reanimate the dead, most notably the head and body of convict George Foster, which were separated hours before at a London prison. Aldini was also a physicist, the professor at Bologna University no less. His research concerned the nature of fire and he even designed lighthouses.

Franz Joseph Gall

Born	March 9, 1758
Birthplace	Tiefenbronn, Germany
Died	August 22, 1828
Importance	Founder of phrenology

Remembered as the father of phrenology, a theory he had developed since childhood, Franz Joseph Gall cuts an eccentric figure from a modern point of view. Although his theories on skull shape and personality were misplaced, they provided the fuel that led to the functional brain maps that we use today. Some of those advances were thanks to Gall's regular anatomical research carried out while working in mental asylums. Gall's phrenology attempted to locate mental faculties in the brain, and this field of research, which looked at the content of the mind, motivation, and decision-making, became known as "psychology."

Karl August Weinhold

Born	October 6, 1782
Birthplace	Meissen, Germany
Died	September 29, 1829
Importance	Studied electrical activity in brain

This German is remembered for replacing a cat's brain with a mixture of metals—and announcing the astonishing (and rather unbelievable) news that the cat was still able to move. However, this is not his chief claim to notoriety. He began as an army doctor before entering Prussian high society as an advisor to the king. One of Weinhold's concerns was overpopulation. His suggestion was to have every unmarried woman infibulated: In other words, have their labia clamped together using removable metal clips. Mercifully, despite Weinhold's assurances that the procedure was as straightforward as giving an injection, the government did not make this policy.

Jan Evangelista Purkyne

Born	December 17, 1787
Birthplace	Libochovice (now Czech Republic)
Died	July 28, 1869
Importance	Discoverer of neurons

Purkyne has many things named after him, and they are not always related. In the context of the brain, there are Purkyne cells, which are large neurons found in the cerebellum and were among the first nerve cells described. Then there are Purkyne fibers, which are involved in the contraction of the heart muscle. There are a whole host of Purkyne eponyms to do with the way the eye sees colors. Purkyne also cataloged the patterns in fingerprints and named the components of blood, and we still use his term "plasma" today. Away from medicine, Purkyne showed an early interest in movie photography.

Bernhard von Gudden

Born	June 7, 1824
Birthplace	Kleve, Germany
Died	June 13, 1886
Importance	Inventor of brain microtome

Von Gudden's contribution to neuroscience was the microtome; however, this is overshadowed by his relationship with King Ludwig II, the "Mad King" of Bavaria. Von Gudden rose to be the preeminent psychiatrist in Bavaria. As a result he was appointed as the king's doctor, partly because it was needed, and partly because the authorities were looking for a way of getting the profligate monarch off the throne. On June 10, 1886, the king was deposed and he accompanied von Gudden to a secluded lakeside residence. No one knows what happened next, but three days later, both the king and his doctor were found dead floating in the lake.

Theodor Schwann

Born	December 7, 1810
Birthplace	Neuss, Germany
Died	January 11, 1882
Importance	Developer of cell theory

A leading figure in the cell theory that underpins biology, Theodor Schwann got his inspiration from studying nerve cells. Cell theory declares that all life, in its great diversity, is based on the small cell unit, of which the nerve cell is just one type. Schwann's work on nerves is remembered by naming a type of glial cell for him. He was also instrumental in revealing the cellular structure of muscles and he discovered the digestive enzyme "pepsin," coined the term "metabolism," and figured out that the yeast that made bread rise was itself a cellular organism.

Paul Broca

Born	June 28, 1824
Birthplace	Sainte-Foy-la-Grande, France
Died	July 9, 1880
Importance	Located speech area in frontal lobe

Paul Broca was something of a scientific rebel. He graduated from university at the age of 16 and was a qualified doctor by 20. After spending his early career as a dermatologist and urologist, in 1848, Broca became a founding figure in a society exploring the theory of evolution. He was sympathetic to Darwin's ideas which circulated in the scientific community long before they became common knowledge in 1859. Broca's attitude, summed up a, "I would rather be a transformed ape than a degenerate son of Adam," brought him into conflict with the church and state. However, his reputation survived and in later life Broca served in the French Senate.

Jean-Martin Charcot

Born	November 29, 1825
Birthplace	Paris, France
Died	August 16, 1893
Importance	Founder of modern neurology

Among a large field of contenders, Jean-Martin Charcot has claimed the title as the "founder of neurology," the medical arm of neuroscience. His founding of a neurology clinic at the Salpêtrière hospital in Paris, the first of its kind anywhere in Europe, had far-reaching effects, not least on the development of psychiatry and psychology.

Charcot's primary research was into hysteria, which was later reclassified by others as a host of mental illnesses, or neuroses. Among his protégés were Tourette, Babinski, and Freud, while Charcot himself is remembered by a host of eponyms including Charcot disease (also known as ALS).

John Hughlings Jackson

Born	April 4, 1835
Birthplace	Green Hammerton, England
Died	October 7, 1911
Importance	Proposed hierarchical organization in brain

The son of a brewer, Jackson qualified as a doctor in London, and after a short stint in Yorkshire returned to work in the capital. He took an interest in epilepsy. This work led to him formulating the concept for which he is remembered: The nervous system was organized into three levels. The lowest level handled primitive functions shared with all

animals, the middle level was associated with voluntary movements, while the upper one controlled "human" faculties. Injury to the higher levels led to the patient degenerating to a more "animal" state, in a process that Jackson compared to a reverse version of evolution.

Henry Maudsley

Born	February 5, 1835
Birthplace	Giggleswick, England
Died	January 23, 1918
Importance	Developed concept of personality disorder

This pioneering English psychiatrist was born in the wilds of Yorkshire, the son of wealthy farm owners. His mother died young and so the young Henry was raised by his aunt, who had him tutored intensively from an early age. He did well at university but was reported to have fallen out with many of his teachers, and perhaps as a result found his ambitions to be a surgeon blocked. After eventually settling on psychiatry, Maudsley took over the running of the private asylum started by his father-in-law. In later life, Maudsley is said to have regretted his choice of career and devoted himself to philosophy and cricket.

Eduard Hitzig

Born	February 6, 1838
Birthplace	Berlin, Germany
Died	August 20, 1907
Importance	Located motor cortex

Eduard Hitzig was a pioneer in using electricity to study the brain, revealing the "motor strip" that controlled voluntary movements. His interest in electrical stimulation began while serving as a medical officer in the Prussian Army at the start of his career. He found that applying electric current to the fractured skulls of wounded soldiers caused involuntary muscle spasms. After his groundbreaking work on dogs carried out with Gustav Fritsch, Hitzig pursued a career in psychiatry in Switzerland before eventually becoming a professor at the University in Halle in what is now Germany.

Gustav Fritsch

Born	March 5, 1838
Birthplace	Cottbus, Germany
Died	June 12, 1927
Importance	Located motor cortex

Gustav Fritsch was the partner of Eduard Hitzig, and the pair are credited with proving the first direct link between a brain region and a specific function. Fritsch's involvement—the experimenters worked in a bedroom at Hitzig's home—led to a position at the University of Berlin. However, lab work was not enough for him. Fritsch was an avid explorer, traveling through southern Africa in the 1860s and visiting the Middle East to watch the transit of Venus in 1874, combined with some archaeology in Egypt. In the 1880s, he dabbled in zoology and became an expert in electric fish.

William James

Born	January 11, 1842
Birthplace	New York City, US
Died	August 26, 1910
Importance	Proposed a theory of emotion

William is not the most famous member of the James family. His brother Henry James was the author of *The Portrait of a Lady, The Turn of the Screw*, and other famous American novels. His father was a notable theologian while his sister is also remembered for her diary. (This relates her battle with illness and hints at an incestuous attraction to William.) The

young William began with ambitions to be a painter but opted for medicine instead. After graduating he became interested in psychology (he suffered mental illness himself). Philosophy, which he taught at Harvard until 1907, was his third career.

Hermann Munk

Born	1839
Birthplace	Posen, Poland
Died	1912
Importance	Located visual cortex

Hermann Munk was a Jewish German born in what is now Poland. His early career at the University of Berlin was in the veterinary school and centered on the life cycle of threadworms. Threadworms can cause blindness and Munk's route into brain science came by studying eyesight in dogs. He found that damage to the occipital lobe led to total blindness. He also discovered "psychic blindness" where a brain-damaged dog could see objects to navigate around them, but was unable to recognize previously known objects by sight. This work provided the first clues into how the visual cortex was involved in visual memory, as well as pure perception.

David Ferrier

Born	January 13, 1843
Birthplace	Woodside, Scotland
Died	March 19,1928
Importance	Revealed location of sensory cortex

Ferrier became interested in the brain and psychology while still at medical school in Aberdeen. He worked with Hermann von Helmholtz in Germany as part of his experimental psychology team, a new field at the time.

Ferrier then found work in London, where he met John Hughlings Jackson who was investigating the links between senses and motor responses. Inspired by Jackson, Ferrier began similar research in Yorkshire. Ferrier worked on live animals and was one of the first people targeted by the antivivisectionist movement.

Camillo Golgi

Born	July 7, 1843
Birthplace	Corteno (now Corteno Golgi), Italy
Died	January 21, 1926
Importance	Developed stain for neuron

Camillo Golgi made quite an impact on neuroscience and biology at large, so much so that his home town of Corteno has since been renamed Corteno Golgi in his honor. Golgi followed his father into medicine but was more focused on research than practicing. He graduated as a doctor in 1865 and developed his staining technique in the early 1870s, revolutionizing the study of neurons. Among many breakthroughs, Golgi helped figure out the structure of the kidney and also discovered the network of tubes that are involved in releasing material from cells, which is now called the Golgi apparatus.

Emil Kraepelin

Born	February 15, 1856
Birthplace	Neustrelitz, Germany
Died	October 7, 1926
Importance	Describes bipolar disorder

Kraepelin is regarded by many as the founder of modern psychiatry and was a pioneering researcher in the use of drugs to influence the brain. Kraepelin believed the origin of psychiatric disease was biological. These ideas are well regarded today, despite being eclipsed by the influence of Sigmund Freud in the 20th century. In 1883, Kraepelin published *Compendium der Psychiatrie* in which he argued for research into the physical causes of mental disease and laid the foundations of a classification system for mental disorders. Kraepelin championed better conditions in asylums and called for treatment, instead of imprisonment for the mentally ill.

Santiago Ramón y Cajal

Born	May 1, 1852
Birthplace	Petilla de Aragón, Spain
Died	October 17, 1934
Importance	Developed neuron doctrine

As a child, Santiago was trouble. He was moved school several times due to his unruly behavior, and devoted his time to painting and athletics, much to the dislike of his parents. He was apprenticed as a barber, and in an attempt to interest him in medicine, his father took him to a graveyard to collect human remains! Santiago enjoyed sketching the bones and began medical studies at Zaragoza. He was enlisted as a medical officer and served in Cuba, only to be struck down by disease. He returned to a career in research and combined the microtome with Golgi's stain to study the growth of nerve cells.

Sigmund Freud

Born	May 6, 1856
Birthplace	Freiberg (now Pribor, Czech Republic)
Died	September 23, 1939
Importance	Developed psychoanalysis

For a man who placed childhood at the heart of his thinking, Freud's home life had its fair share of insecurity. His father, Jakob, had two sons by an earlier marriage, and his second wife, Freud's mother, was considerably younger. The young Sigmund played mostly with his nephew, John, who moved away when Sigmund was four. By the age of nine, Sigmund had six more siblings. He was devoted to his mother, but had a distant relationship with his father. In 1886, Freud began private practice in Vienna, where he stayed until 1938 when, as a Jew, he had to flee the Nazis. He moved to London where he died the following year.

Eugen Bleuler

Born	April 30, 1857
Birthplace	Zollikon, Switzerland
Died	July 15, 1939
Importance	Described schizophrenia

A leading figure in the understanding of schizophrenia, Bleuler, the son of a farmer, studied under both Jean-Martin Charcot in Paris and Bernhard von Gudden in Munich. He established a practice in Zurich where he was made director. Bleuler thought that the causes of mental illness lay in the unconscious, as did Freud. He performed a self-analysis with Freud, beginning in 1905. However, he became disillusioned by this idea and began to look for physical causes of mental disorders. This led to his work on schizophrenia, and Bleuler also explored the link between alcohol and mental problems.

Joseph Babinski

Born	November 17, 1857
Birthplace	Paris, France
Died	October 29, 1932
Importance	Discovered plantar reflex

This French-Polish physician is best known for Babinski's sign, where the big toe curls upward when the sole of the foot is stimulated. The sign is a natural reflex in newborns (and while sleeping), but it is the opposite response to the normal action of an adult foot. Babinski made the link between the sign and neurological problems while working in a Paris hospital in 1896. The sign is a diagnostic test today, and leads into further investigations of the spinal cord and brain. Babinski's early career was as an assistant to Jean-Martin Charcot (it is he who is holding the "hysterical" woman on page 63).

Georges Gilles de la Tourette

Born	October 30, 1857
Birthplace	Saint-Gervais-les-Trois-Clochers, France
Died	May 26, 1904
Importance	Described Tourette's syndrome

Gilles de la Tourette studied medicine at Poitiers. After qualifying he became the student and then personal assistant of Jean-Martin Charcot in Paris. Tourette was a leading advocate in taking mental illness into account during criminal trials. In 1893, a female patient shot Tourette in the head, claiming he had hypnotized her and caused her symptoms. He survived but the experience, coupled with the loss of his young son, made Tourette suffer something akin to bipolar disorder. For the last two years of his life, he was confined to a psychiatric hospital.

Charles Scott Sherrington

Born	November 27, 1857
Birthplace	London, England
Died	March 4, 1952
Importance	Discovered the synapse

Sherrington's heritage is unclear. He took his name from James Sherrington but he is believed to be the illegitimate son of Caleb Rose. Both these men were doctors and Sherrington became a surgeon in 1878, developing an interest in physiology. However, he undertook a lengthy sabbatical, traveling through Europe meeting famous names in the field of neurology before returning to England to begin his academic career. Sherrington won the Nobel Prize in 1932 for his work on the reflex arc. This is the nervous pathway between touch sensors and muscles that runs via the spinal cord and operates independently of the brain.

Alois Alzheimer

Born	June 14, 1864
Birthplace	Marktbreit, Germany
Died	December 19, 1915
Importance	Described main form of dementia

Alois Alzheimer's name is all too familiar in the developed world, where 3 percent of people over 65 suffer from the form of dementia associated with him. Alzheimer described the disease in the early 1900s and the promotion of this work by Emil Kraepelin led to it being named after him. Alzheimer spent most of his working life running a clinic in Munich (although he met his first Alzheimer's patient in Frankfurt). In 1912, he was offered a position as professor in Breslau (in what is now Poland). He fell ill on the train traveling to his new home, suffering from an infection. He never really recovered and died three years later.

Carl Jung

Born	July 26, 1875
Birthplace	Kesswil, Switzerland
Died	June 6, 1961
Importance	Develops analytical psychology

Carl Jung's early years were touched by tragedy. He was the only surviving son of a Swiss pastor, and his mother suffered from depression and delusions, spending large parts of Carl's childhood in hospital. As a child, Carl believed he had two personas: A child and the other an old, well-respected man from a past era. He had a strong relationship with his father, but felt let down by his mother, leading to a certain misogyny—and he was a philanderer in adulthood. Jungian psychology, which proposes that personalities are the product of archetypes at work in the subconscious, was the result of Jung's mixture of mystical and emotional experiences.

Otto Loewi

Born	June 3, 1873
Birthplace	Frankfurt, Germany
Died	December 25, 1961
Importance	Discovered neurotransmitter

Loewi had intended to care for the sick, working in the hospitals of Frankfurt. However, in the late 1890s, he became frustrated that so many of his patients were suffering from diseases that had no treatments, and so Loewi switched to a career in research. He became friends with Englishman Henry Dale, who discovered acetycholine, and it was this chemical that Loewi showed was involved in transferring nervous signals to tissues. In 1936, the friends were jointly awarded the Nobel Prize. Two year later Loewi, a Jew, was forced to flee Germany. Penniless, he was taken in by Dale in London and then moved to the United States.

Nathaniel Kleitman

Born	April 26, 1895
Birthplace	Chisinau (now Moldova)
Died	August 13, 1999
Importance	Founder of sleep research

Nathaniel Kleitman life's work was done in Chicago. The Jewish Kleitman had emigrated to escape persecution, eventually arriving in New York in 1915. Within ten years he was working as part of the University of Chicago, having already developed a specialism in sleep. His interests in sleep came from an interest in consciousness, and he chose to find out more by studying the unconsciousness of sleep. The research was funded in part by the Wander Company, who had an interest in sleep. Their malt drink Ovaltine was marketed as a cure for insomnia.

Gray Walter

Born	February 19, 1910
Birthplace	Kansas City, Missouri, USA
Died	May 6, 1977
Importance	Leading figure in artificial intelligence

American by birth, Walter moved to London as a child. In 1939 he was offered a post at the Burden Neurological Institute in Bristol, England, and stayed there until 1970. He used ECGs to map the locations of activity in the brain, and was even able to use this system to locate tumors. He also discovered the "readiness potential" that preceded conscious actions. His greatest achievements were the "turtle" robots, which he used to explore the limits of nerve cell networks. In 1970, Walter had an accident on his scooter and suffered a debilitating brain injury, from which he never fully recovered.

Andrew Huxley

Born	November 22, 1917
Birthplace	London, England
Died	May 30, 2012
Importance	Discovered action potential

Andrew was born into the famous Huxley family—his half brothers were Aldous (author) and Julian (biologist and conservationist), while his grandfather Thomas Huxley was "Darwin's Bulldog," the chief advocate of the theory of evolution in the 1860s. Andrew did much to honor the name with his Nobel Prize-winning work alongside Alan Hodgkin. This work revealed the action potential that carries nerve signals as electric pulses. Huxley went on to collaborate with German physiologist Rolf Niedergerke, to figure out how muscles contracted after receiving their electronic signal.

Alan Lloyd Hodgkin

Born	February 5, 1914
Birthplace	Banbury, England
Died	December 20, 1998
Importance	Discovered action potential

Hodgkin was selected to work on the top-secret radar defense systems being developed in Britain to counter the threat of invasion during World War II. He flew the first radar-enabled test flight in 1941. After the war, Hodgkin joined Cambridge University where he continued his collaboration with Andrew Huxley, begun in 1935. By 1952, they were ready to publish results, which showed how a pulse of electrical potential moved along the axon of a neuron. This action potential was the source of all the brain and body's electrical activity. The discovery won them the Nobel Prize for Physiology and Medicine in 1963.

Eric Kandel

Born	November 7, 1929
Birthplace	Vienna, Austria
Died	—
Importance	Revealed chemistry of memory

Kandel and his family left their home after Austria was annexed by Nazi Germany in 1938. Their new home was to be Brooklyn, New York. Kandel's first degree, from Harvard, was in history and literature, when he explored the rise of National Socialism (Nazism). However, while at Harvard Kandel became interested in the work of B.F. Skinner, who drew a strict distinction between psychology and neuroscience. Kandel began to study memory in an attempt to understand any link between the two. In the 1960s and 1970s Kandel discovered the chemical basis of memory, work for which he received the Nobel Prize in 2000.

BIBLIOGRAPHY AND OTHER RESOURCES

Al-Chalabi, Ammar. *The Brain: A Beginner's Guide.* Oneworld Publications, 2008.

Amthor, Frank. *Neuroscience for Dummies.* Wiley, 2012.

Blakemore, Colin. *The Mind Machine.* BBC Books, 1994.

Buzsaki, Gyorgy. *Rhythms of the Brain.* Oxford University Press, 2006.

Damasio, Antonio R. *Descartes' Error: Emotion, Reason, and the Human Brain.* Penguin, 2005.

Doidge, Norman. *The Brain that Changes Itself: Stories of Personal Triumph from the Frontiers of Brain Science.* Viking, 2007.

Finger, Stanley. *Origins of Neuroscience.* Oxford University Press, 1994

LeDoux, Joseph. *Synaptic Self: How Our Brains Become Who We Are.* Viking, 2002.

Rose, Steven. *The Future of the Brain: The Promise and Perils of Tomorrow's Neuroscience.* Oxford University Press, 2006.

Sacks, Oliver. *The Man Who Mistook His Wife for a Hat and Other Clinical Tales.* Touchstone, 1985.

Societies

Belgian Brain Council (www.belgianbraincouncil.be)

Brazilian Society for Neuroscience and Behavior (www.sbnec.org.br)

British Neuroscience Association (bna.org.uk)

Canadian Association for Neuroscience (can-acn.org)

Chinese Neuroscience Society (www.csn.org.cn)

Dutch Neurofederation (neurofederatie.nl)

French Neuroscience Society (fens.org)

German Neuroscience Society (nwg.glia.mdc-berlin.de/de)

Indian Academy of Neurosciences (neuroscienceacademy.org.in)

Italian Society for Neuroscience (www.sins.it)

Japan Neuroscience Society (www.jnss.org)

Korean Society of Brain and Neuroscience (www.ksbns.org)

Polish Neuroscience Society (ptbun.org.pl)

Russian Neuroscience Society (neuroscience.ru)

Society for Neuroscience, USA (www.sfn.org)

Spanish Society of Neuroscience (senc.es)

Swedish Society for Neuroscience (www.ssfn.se)

International Organizations

Australasian Neuroscience Society (ans.org.au)

European Brain and Behaviour Society (ebbs-science.org)

Federation of European Neuroscience Societies (fens.org)

Federation of Neuroscience Societies in Latin America, the Caribbean, and the Iberian Peninsula (falan-ibrolarc.org)

International Brain Research Organization (ibro.info)

Organization for Human Brain Mapping (humanbrainmapping.org)

Websites

www.brainfacts.org

Dana Brainweb: dana.org/brainweb

www.humanbrainproject.eu

Massachusetts Institute of Technology Open Courseware: ocw.mit.edu/courses/#brain-and-cognitive-sciences

Whole Brain Atlas: med.harvard.edu/AANLIB/home.html

Apps

3D Brain

3D Neuron Anatomy

Brain Lab

Brain & Nerves

Brain Tutor 3D

Netter's Neuroscience Flash Cards

Neuroknowledge

Places to Visit

Alois Alzheimer's Birthplace, Marktbreit, Germany (www.alzheimer-haus.de)

Arizona Science Center, Phoenix, Arizona, USA (azscience.org)

Brain Museum, Lima, Peru (www.icn.minsa.gob.pe)

Cajal Museum, Cajal Institute, Madrid, Spain (www.cajal.csic.es)

Camillo Golgi Museum, Corteno Golgi (www.museogolgi.it)

Cushing Center, Yale University, USA (library.medicine.yale.edu/cushingcenter)

Freud Museum, London, UK (freud.org.uk)

Gordon Museum of Pathology, King's College London, UK (kcl.ac.uk/gordon)

Marian Koshland Science Museum, Washington DC, USA (koshland-science-museum.org)

Museum of the History of Science, Oxford, UK (www.mhs.ox.ac.uk)

Museum of the Spanish Society of Neurology, Barcelona, Spain (mah.sen.es)

Mütter Museum, College of Physicians of Philadelphia, USA (muttermuseum.org)

National Museum of Health and Medicine, Silver Spring, Maryland, USA (www.medicalmuseum.mil)

The Wellcome Collection, London, UK (wellcomecollection.org)

Archives

Jean-Martin Charcot correspondence, United States National Library of Medicine, Bethesda, Maryland, USA (www.nlm.nih.gov)

Jean-Martin Charcot Library, School of Neurology, Hôpital Salpetrière, Paris, France (www.aphp.fr)

Alan Lloyd Hodgkin papers, Trinity College Library, University of Cambridge, UK (www.trin.cam.ac.uk)

John Hughlings Jackson papers, Queen Square Library, London, UK (neuroarchives@ucl.ac.uk)

Carl Jung papers, Swiss National Library, Bern, Switzerland (nb.admin.ch)

INDEX

A

acetylcholine 90
action potential 104
acupuncture 13
adrenaline 90, 92, 122
affective disorders 53
Alcmaeon 15, 17
Aldini, Giovanni 35, 132
Algeria 19
Al-Haytham 20
Allegory of the Cave, Plato's 16
alpha waves 63, 107, 128
al-Razi 27
Alzheimer, Alois 78–79, 138
Alzheimer's disease 78–79
anatomy 6, 8, 15, 18, 21, 23–25, 34, 40, 45, 50–51, 53, 58, 61, 86, 97, 116, 130–131
An Essay Concerning Human Understanding 33
anesthetics 42, 43
angiography 95
animal brains 7, 62, 126
animal spirits 19, 25, 29, 31, 34, 54, 88
anxiety 53, 70, 103
apes 50–51, 55, 57, 100, 126
aphasia 48, 69
apoplexy 26–27
apraxia 78
archetypes, Jungian 71, 138
Aristotle 15, 17–18, 20, 88, 130
artificial intelligence (AI) 117
Aserinsky, Eugene 106
Asperger, Hans 96
astrocyctes 47
auditory canal 45
auditory cortex 99
autism 96
autonomic nervous system 76, 92, 120, 122
Avicenna 7, 20, 29, 31, 49, 67, 131
axon 40–41, 47, 75, 90, 104–105, 139

B

Beddoes, Thomas 42
behaviorism, radical 100, 103
Bell–Magendie Law 39
Berger, Hans 62–63
Berkeley, George 33
beta waves 63
Binet, Alfred 86–87
binocular vision 34
bipolar disorder 77, 136–137
black reaction 60
Bleuler, Eugen 82–83, 96, 137
blindness 6, 65, 80, 135
blood pressure 81, 92, 130
blood sugar 92
Boerhaave, Herman 64–65
brain stem 8–9
brain surgery 10
Broca, Paul 48, 51, 68–69, 81–82, 101, 133
Broca's area 48, 69, 81–82
Brodmann, Korbinian 80–81

C

camera obscura 21, 58
carotid artery 19, 27, 95
cell membrane 104
cell theory 41, 54, 133
central nervous system 120
cephalopods 60
cerebellum 8–9, 88
cerebral dominance 68
cerebral hemispheres 65
cerebrum 8
Chalmers, David 115
Charcot, Jean-Martin 7, 38, 63, 70, 134, 137
China 13
chloroform 42–43
chorea 22, 26, 32, 38
choroid 58
Circle of Willis 30
Civil War, American 56
cochlea 45, 93, 99
code, neural 130
Cogito ergo sum 28
cognitive behavioral therapy 103
coma 109
Commodus, Emperor 18
common sense 6, 31, 67
cone, eye cell 59
conjunctiva 58–59
consciousness 16, 18, 20, 28–29, 33, 43–44, 61, 85, 107, 114–115, 128–129, 138
constitutional psychology 97
coprolalia 66
cornea 58
corpus callosum 8, 98–99
Corti, Alfonso 45–46, 93
cramp, muscle 105
cranial nerves 45, 120–121
cranium 8
Creation of Adam 24
Creutzfeldt-Jakob disease 79
criminal trials, neuroscience 116
crustaceans 60
crying 127
curare 42
cytoarchitectonics 80

D

Dale, Henry 90
dancing mania 22
Darwin, Charles 53, 57, 67, 86, 133, 139
deafness 6
De humani corporis fabrica 25
dementia 78–79, 82, 138
Democritus 16, 17
dendrites 40–41, 72, 74–75
dermatome 73
Descartes, René 28–30, 100, 114
Deter, Auguste 78
dissociative identity disorder 83
Djoser, pharaoh 12
dopamine 38
Down's syndrome 51
dreaming 129
dualism 29
Duchenne de Boulogne 6, 52
dyslexia 80

E

ear 23, 37, 45, 93, 99, 104
eardrum 45
ECG 62
echolalia 66
echopraxia 66
ectomorph 97
EEG 62, 104, 106–107, 110
effector 55
ego 70–71
Egyptians, ancient 12
Einstein's IQ 87
electrical stimulation 7, 52, 55
electricity 34–35, 47, 52, 63, 104–105, 117, 132–134
electrocardiograph 62
electroconvulsive therapy 94
electroencephalograph 62
elements, classical 13–14, 111, 132
Empedocles 17
endomorph 97
engram 8, 108
Epicurus 15
epilepsy 62–63, 84–85, 98, 134
episodic memory 108
equipotentiality 91
ether 43
Eustachio, Bartholomeo 45
Evil Eye, the 13
extramission theory 15, 34

F

Fallopius, Gabriel 45
Faraday, Michael 43
Ferrier, David 55, 73, 99, 135
"fixed ideas" 95
Flying Man, thought experiment 20
forebrain 8–9, 52, 92
fovea 59
Frankenstein 35
Franklin, Benjamin 94

Franz, Shepherd 36, 48, 59, 61, 64, 91, 108, 132
free will 100, 115
Freudianism 70
Freud, Sigmund 60, 70–71, 103, 134, 136–137
Fritsch, Gustav 52, 55, 134–135
frontal lobe 9
functional anatomy 9, 80–81
Furies, Greek mythology 14

G

Gage, Phineas 44, 68
Galen 18–19, 22, 25, 27, 34, 42, 46, 58, 76, 88, 130
Galen's nerve 18
Gall, Franz Joseph 36
Galton, Francis 86, 97
Galvani, Luigi 34
gamma waves 63
Gestalt 89
Gettysburg, Battle of 56
gladiators 18
Glasgow Coma Scale 109
glial cells 47
Gliddon, George 50–51
God Helmet 113
Golgi apparatus 60, 136
Golgi, Camillo 60, 72, 74–75, 136
gray matter 55
Gray Walter, William 102, 139
Greek mythology 14
Gudden, Bernhard von 61, 74, 133–137
gyrus 9, 31, 55, 101, 119

H

hallucinations 23, 83
handedness 126
hard questions of consciousness 115
Hawking, Stephen 117
head injuries 6
hearing, theories 93
Hebb, Donald 108
Helmholtz, Hermann von 93, 135
hemispacial neglect 98
Henning's prism 46
Herophilus 15
hippocampus 8, 79, 101, 128
Hippocrates 13–15, 18, 22, 26–27, 42, 130
Hitzig, Eduard 52, 134–135
Hodgkin, Alan 104
holism 72–73
Holy Trinity 19
Homer 17
hominid 10
homunculus, motor and sensory 118
Hooke, Robert 40
Human Connectome Project 8
humoral theory 14
Huntington's disease 26
Huxley, Aldous 97
Huxley, Andrew 104, 139
hypnotism 63
hypothalamus 8, 34, 92
hysteria 23, 63, 134

I

Imhotep 12–13
immaterialism 33
imponderables 126, 128, 130
inattention 99
infrasound 113
intelligence 13, 31, 51, 86–87, 98, 102, 117, 127, 131, 139
intentionality 61
interneurone 55
intromission theory 15
IQ 86–87
iris 58, 122
"I think, therefore I am." 28
Ivy League nude photographs 97

J

Jackson, John Hughlings 38, 69, 78, 101, 134–135
James–Lange Theory of Emotion 67
Jekyll and Hyde 68
jugular vein 45
jumping Frenchmen of Maine 66
Jung, Carl 71, 138

K

Kandel, Eric 108, 139
Katz, Bernard 104
Kleitman, Nathaniel 106–107, 138
Kraepelin, Emil 77, 82, 136, 138

L

Lascaux Caves 11
Lashley, Karl 91, 108
lateral geniculate nucleus 64–65
lateral sulcus 9
laughing gas 42–43
Leonardo da Vinci 23, 24
leucotomy 95
limbic system 92, 101
Line of Gennari 65
Linnaeus, Carl 50
lithium 77
lobotomy 95
localization, cerebral 9, 36, 39, 80, 82, 131
Locke, John 33, 111
Loewi, Otto 90, 138
Long, Crawford 43
Ludwig II, "Mad King" of Bavaria 61, 133

M

machina speculatrix 102
MacLean, Paul 101
Magnetic Resonance Imaging (MRI) 112
Magnus, Albertus 19, 131
mania 22–23, 53, 77, 94
mass action 91
Maudsley, Henry 53, 57, 134
medical imaging 7, 81
medulla oblongata 8
memory recall 129
memory trace 108
meninges 8
Mensa 87
mesomorph 97
Michelangelo 24
microglia 47
microscope 7, 40, 49, 58–59, 74–75
microtome 61, 74, 133, 136
midbrain 8
mitrochondrian 90
Moniz, Egas 95
mood disorder 53
motor cortex 32, 55–56, 81, 85, 118–119, 134–135
motor nerve 39, 105
mummifications 12
Munk, Hermann 65, 73, 135

N

narcolepsy 64
Nemesius 19
nervous system 6–7, 18, 23, 26, 39–40, 52, 57, 63, 76–77, 92, 94, 120, 122, 134
neural nets, computing 117
neural reticulum 54
neurology 6, 7, 45, 116
neuron doctrine 54, 74, 104, 136
neurons 6, 47, 55, 60, 74–75, 77, 105, 108, 129–130, 133, 136
neuroscience 6, 18–20, 25, 28, 33–34, 36, 44, 48, 56, 61, 68, 72, 74, 76, 86, 95, 104, 108, 110–111, 113, 116, 126, 129, 131–130, 133–134, 136, 139
neurotransmitters 90
nitrous oxide 42
Nobel Prize 75, 90, 108, 137–139
nose 7–8, 12, 19, 34, 42, 46, 49
Nott, Josiah 50–51
nucleus 40, 64–65, 101

O

occipital lobe 9, 65, 88, 135

olfaction 46
On the Origin of Species 57
opium 42
optic chiasm 34
optic nerve 59, 121
optical theory of vision 21
organelles 60

P
paralysis 6, 14, 23, 38, 88, 106
paranoia 57, 70, 83, 95
parapsychology 113
parasympathetic, nervous system 76, 122–123
Paré, Ambroise 56
Parfit, Derek 111
parietal bone 11
parietal lobe 9
Parkinson's disease 38
peripheral nervous system 120
personality 6, 22, 26, 44, 63, 68, 83, 97, 116, 132, 134
personality disorders 116
phantom limbs 56
phenobarbital 84–85
phrenology 36, 64
pituitary gland 8
Plato 16–17, 20, 130
pons 8, 122
Positron Emission Tomography (PET) 110
postcentral gyrus 9
postures, coma 109
poverty of stimulus 131
precentral gyrus 9
prediction center 131
Prince, Morton 82, 84
principle of dissolution 78
prion 79
psychic blindness 135
psychic energy 62
psychoanalysis 70
pupil, eye 17, 21, 58, 76, 130
Purkinje, Jan Evangelista 40–41

Q
qualia 114–115

R
racism and neuroscience 50
rage 14, 44, 57, 92
Ramón y Cajal, Santiago 41, 75
receptor 55, 90
reflex action 55
refrigerator mothers 96
REM sleep 106–107, 129
Renaissance 22, 24, 50
reptilian brain 101
retina 21, 58–59
rhodopsin 59
rod, eye cell 59, 93
Romans, ancient 18, 32, 58, 76, 80, 131
Ryle, Gilbert 100

S
Salem witch trials 22
Salpêtrière Hospital 70
schizophrenia 82–83, 94, 96, 137
Schwann, Theodor 41, 47, 133
sclera 58
sea slugs 108
sensory nerve 39, 121
sham rage 92
Sheldon, William 97
Shelley, Mary 35
Sherrington, Charles 74–75, 90, 137
Sistine Chapel 24
Skinner, B.F. 100, 102–103, 139
skull 8, 10–12, 14, 23, 25, 35–37, 44, 48, 50–52, 65, 86, 132
sleep 17, 43, 63, 64, 72, 81, 106–107, 109, 115, 128–129, 138
sleep cycle 106
sleep deprivation 64, 72
sleep paralysis 106
smells 46
somatic nerves 120
somatosensory cortex 55, 118–119
somatotype 97
Sömmerring, Samuel von 50, 58
soul, tripartate 16
speech 6, 27, 36, 48, 66, 68–69, 80–82, 91, 109, 131, 133
spinal cord 8-9, 120
split personality 83
Standford–Binet Scale 87
St. Anthony's Fire 23, 90
St. Augustine of Hippo 19
Stone Age 10–11
striate cortex 65
striatum 31, 85
stroke 26–27, 48, 65, 69, 98, 132
St. Vitus's Dance 32
substantia nigra 38
sulcus 9, 31
supercomputers 117
superego 71
Sushruta 13
Swedenborg, Emanuel 81
Sylvian fissure 9
sympathetic nervous system 76, 122–123, 133

T
tabula rasa 33
talking cure 70, 103
Tan, patient 48
taste buds 49
teleportation 111
temporal lobe 9
theories of vision 15
theory of mind 96, 127
Theseus's Ship 111
time, neural basis of 128
touch receptors 73
Tourette, Georges Gilles de la 66, 134, 137
Tourette's syndrome 66, 137
trepanning 10–11
"turtle" robots 102, 139

V
vagus nerve 76, 120
van Gogh, Vincent 77, 116
ventricle 8, 19, 31, 88
Vesalius, Andreas 25, 45–46, 131
vision 9, 15, 21, 28, 34, 59, 64–65, 114, 131
visual cortex 64
voxels 129

W
Weinhold, Karl August 35, 132
Wepfer, Johann Jakob 27, 132
Wernicke's area 69, 80–81
white matter 55
Willis, Thomas 6, 30–31, 76, 85
witches 22
World War I 88, 94
Wounded Man 11

Y
yin and yang 13
Young Simpson, James 43

Cataloging-in-Publication Data has been applied for and may be obtained from the Library of Congress.

ISBN 978-0-9853230-8-0

Series Concept and Direction: Jeanette Limondjian
Design: Bradbury and Williams
Editor: Meredith MacArdle
Proofreader: Marion Dent
Picture Research: Clare Newman
Consultant: Dr. Kelley Remole
Cover Design: Igor Satanovsky

SHELTER HARBOR PRESS
603 West 115th Street Suite 163
New York, New York 10025

For sales in the U.S. and Canada, please contact
info@shelterharborpress.com

For sales in the UK and Europe, please contact
info@worthpress.co.uk

Printed and bound in China by Imago.

10 9 8 7 6 5 4 3 2 1

PICTURE CREDITS

BOOK

Alamy: Age Fotostock 71tr; Paul Bevitt 10cbr; Scott Camazin 30br; Cini Classico 116crt; Classic Stock 116c; Gianni Dagli Orti/The Art Archive 11tr; Everett Collection 22br; Peter Horree 16bl; Chris Howes/Wild Places 11b; Interfoto 26, 39tc; MEPL 14tr, 18bl, 32bl, 34br; PBL Collection 35tr; Pictorial Press Ltd. 33b, 68cr; Prisma Archivo endpapers; **Corbis:** Bettmann 76, 103; Everett Kennedy Brown 111tl; Hulton Deutsch Collection 86br, 139tr; Louis Psihoyos 138br; Ted Stershinsky 95tr; George Tatge 17; **FLPA:** Mitsuaki Iwago/Minden Pictures 100br; **Getty Images:** Ed Reschke/Photolibrary 60; SSPL 51; **Library of Congress:** 56; **Mary Evans Picture Library:** 4, 12bl, 23cr, 40tr, 52cr, 61tl, 69, 70tr, 79b, 84tr, 84bl, 92tl, 133tr, 134tr, 138tr, 139bl; **SCETI:** Edgar Fahs Smith Collection 18tr; **Shutterstock:** 2–3, 3Dme Creative Studio 31crt; Albund 68bl; Alexilus 59b, 101tl; Alila Medical Media 48cr, 55tr, 65, 84br; Anastasios71 16br, 130bl; Animus81 126b; Artcasta 98cr; Stephanie Bidouze 123; Bike Rider, London 96; Stephan Bormotor 115; Browyn Photo 129b; Vitor Costa 122cl; Design Villa 38bl; Designua 73br, 88tr, 90bl; Goran Djukanovic 101cr; Duco59us 55br; B. Erne 37br; Eveleen 9t; Everett Historical 88cl 131bl, 131br, 136br; Juan Gaertner 54; Johanna Goodyear 122tl; Elsa Hoffmann 99b; Iculig 91tl; Lyricsaima 85tr; Marcos Mesa Sam Wordley 106; Maridav 105tl; Eugenio Marongiu 77br; Martchan 120cl; Sandra Matic 124; Neveshkin Mikolay 14bl; Mistery 128; Mopic 75; Morphart Creation 34tl, 42tl, 93; Dragana Gerasi Mosk 126t; Hein Nouwens 45bl; Tyler Olson 91br; Orlandin 108bl; Amawasri Pakdara 120tl; Photo Fun 23tl; Reinette Graphics 35br; Jamie Roach 120bl; Arun Roisri 33tr; Frederico Rostagno 12tr; Science Pics 98tl; Takito 39br; Dietmar Temps 108tl, 139br; Tommistock 125b; Udaix 92tl; Taras Ver Khovynets 67br; Vitalez 113t; Wallenrock 125tr; Rinat Zevriyev 38tr; zprecech 133tl; **Science Photo Library:** 117b; D. Van Bucher 107; Victor De Schwanbery 102; Sam Flak 100tl; Spencer Grant 72/73; Jacopin 105br; Francis Leroy/Biocosmos 104; Living Art Enterprises 95bl; Medical Images/Universal Images Group 119t, 119b; Afred Pasieka 111br; **Science & Society Picture Library:** Science Museum 61br, 139tl; **Thinkstock:** Jemal Countless/Getty Images News 117tr; Dorling Kindersley 49b; J. Falcetti, J/iStock 40bl, 47b; Feel Life/iStock 7b, 112; Fortish/iStock 79cr; Fuse 10cl; Georgios Kollidas 28c; Alex Luengo/iStock 59t; Andreas Odersky/iStock 85cr; Photos.com 2–3; 22tl, 24, 25bt, 27b, 36tr, 36bl, 37bl, 39tl, 43, 48tr, 131tr, 132bl, 133br; Radio Moscow/iStock 129t; Mark Strozier/iStock 46br, 49tr; Boris Urunlu/iStock 72tl; Wander Luster/iStock 48bl; Wenht/iStock 5, 11tr; Matthew Zinder/iStock 42bt; **Topfoto:** Fortean Blackmore 113b; The Granger Collection 97bl; **U.S. National Library of Medicine:** 20, 25tl, 25tc, 27tr, 31tl, 47tr, 67tl, 71bl, 73trt, 73trc, 73trb, 82bl, 134br, 135tr, 135bl, 137tl, 138tl; **U.S. Government:** 66br; **Wellcome Library, London:** 6tl, 19, 28b, 35tl, 45tl, 53br, 55tl, 57cl, 57cr, 57b, 58tr, 58bl, 66bl, 70bl, 74tr, 74cl, 74cbl, 77tl, 86tr, 90tr, 130tr, 130br, 132tl, 132tr, 134tl, 134bl, 135br, 136tl, 137tr, 137bl, 137br, 138bl; **Wikipedia:** Van Horn, Irima, Torgerson, Chambers, Kikinis 44bl; 6cr, 6bl, 7t, 13tr, 13bl, 15, 21cr, 21b, 29, 30bl, 31crb, 32tr, 39cl, 41b, 44tr, 50tr, 50bl, 52b, 62bl, 62br, 63, 78, 79t, 81b, 82t, 89, 94, 97tl, 116crc, 116crb, 127, 131tl, 135tl, 136bl; **Roy Williams:** 83b, 110b, 114; **Woodman Design:** illustrations/diagrams 8, 9, 31, 38, 39, 40, 47, 48, 55, 79, 81, 84, 85, 89, 90, 92, 97, 98, 101, 104, 105, 119, 120, 121, 123.

TIMELINES

Alamy: A. F. Archive; Ancient Art & Architecture; Coconut Aviation; Mihailo Maricic; Pictorial Press Ltd.; World History Archives; **Corbis:** Bettmann; **Shutterstock:** H. T. Brandon; Chameleons Eye; Decade 3D; Everett Historical; Iloria Ignatora; Jorisvo; Sebastian Kaulitzki; Kletr; Marc Pagani Photography; Morphart Creation; Buelikova Oksana; Ken Tannenbaum; M. Tiara; **Science Photo Library:** Living Art Enterprises; **Science & Society Picture Library:** Science Museum; **Thinkstock:** Daniel Beehulak/Getty Images News; Dell 640/iStock; Janka Dharmasena/iStock; Digital Vision; Judy Dillon; Dorling Kindersley; Aos Fulcanelli/iStock; Georgios Art/iStock; Matt Gibson; Photos.com; Sculpies/iStock; I. V. Serg/iStock; Tirtix/iStock; Wenht/iStock

Publisher's Note: Every effort has been made to trace copyright holders and seek permission to use illustrative material. The publishers wish to apologize for any inadvertent errors or omissions and would be glad to rectify these in future editions.

NATIONAL UNIVERSITY LIBRARY
3 1786 10303 5868